Recent Advances in Pyrolysis

Edited by Hassan Al-Haj Ibrahim

Published in London, United Kingdom

IntechOpen

Supporting open minds since 2005

Recent Advances in Pyrolysis
http://dx.doi.org/10.5772/intechopen.77528
Edited by Hassan Al-Haj Ibrahim

Contributors
Abdelrahman Rabie, Marwa Aboelela, Krishna Yadav, Sheeja Jagadevan, Gartzen Lopez, Itsaso Barbarias, Aitor Arregi, Maite Artetxe, Laura Santamaría, Maria Cortazar, Maider Amutio, Javier Bilbao, Martin Olazar, Udaya Ragula, Sriram Devanathan, Sindhu Subramanian, Ratnawulan Ratnawulan, Ahmad Fauzi, Hassan Al- Haj Ibrahim

First published in London, United Kingdom, 2020 by IntechOpen
IntechOpen is the global imprint of INTECHOPEN LIMITED, registered in England and Wales, registration number: 11086078, 7th floor, 10 Lower Thames Street, London, EC3R 6AF, United Kingdom
Printed in Croatia

British Library Cataloguing-in-Publication Data
A catalogue record for this book is available from the British Library

Additional hard and PDF copies can be obtained from orders@intechopen.com

Recent Advances in Pyrolysis
Edited by Hassan Al-Haj Ibrahim
p. cm.
Print ISBN 978-1-78984-063-6
Online ISBN 978-1-78984-064-3
eBook (PDF) ISBN 978-1-78984-942-4

For EU product safety concerns:
IN TECH d.o.o., Prolaz Marije Krucifikse Kozulić 3, 51000 Rijeka, Croatia, info@intechopen.com or visit our website at intechopen.com.

Meet the editor

Dr. Hassan Al-Haj Ibrahim obtained a PhD in Fuel Engineering from Leeds University, England, in 1973. From 1970 until 1974 he was a Research Fellow at the British Coke Research Association (Leeds University, England) after which he worked at Homs Oil Refinery, Syria (1974–1975). In 1980, he spent an academic year at the Technical University of Aachen, and in 1990 he was awarded a Fulbright scholarship and spent an academic year at the University of Pittsburgh. Dr. Al-Haj Ibrahim served as a professor of Petroleum Refining at Al-Baath University, Syria, for forty-two years and has been its Director of Quality Assurance since 2003. In 2017, he joined the Arab University for Science and Technology in Syria as Director of Quality Assurance and a professor in the Chemical Engineering Department.

His publications include research papers on chemical engineering and quality assurance in education, literature and general culture. He has also written several textbooks on petroleum refining, metallurgy, sources of energy, and a dictionary of petroleum terms.

Contents

Preface

Pyrolysis is an irreversible thermochemical treatment process of materials at elevated temperatures in an inert atmosphere. Pyrolysis is used heavily in the chemical industry to produce many forms of carbon and other chemicals from petroleum, coal, wood, oil shale, biomass or organic waste materials. It is the basis of several methods for producing fuel from biomass. Pyrolysis also is the process of conversion of buried organic matter into fossil fuels.

Many factors contribute to making pyrolysis an attractive and possibly an essential process for converting biomass into bio-fuel or bio-energy. During the last fifty years or so, and almost within living memory, man has witnessed an almost imperceptible change in the energy situation from surplus and cheap availability to scarcity. Renewable sources of energy have to be found in order to meet the continuously rising demand and decreasing energy reserves. Waste products and other low-quality sources of energy and bottom-of-the-barrel residues may no longer be discarded, and pyrolysis is one of the most effective processes for treating and upgrading waste products.

Books and learned articles on pyrolysis are not few in number, but a new book on pyrolysis is always called for in view of the continuous developments and new processes in this important field of the chemical industry. The chapters in this book cover a wide area of this important topic. The introductory chapter in particular gives a brief but comprehensive account of the various pyrolysis processes, products and materials. The other chapters of the book deal with specific processes including, among others, fast pyrolysis and pyrolysis of lignin and biomass materials such as waste tyres and waste plastics. Other topics discussed include synthesis of different pyrolysis products including bio-oil and biochar and modelling and optimization of pyrolysis products.

Readers of this book may not find an exhaustive or an encyclopaedic account of all pyrolysis processes, products and materials, but they will find many interesting and relevant topics related to different aspects of pyrolysis discussed by eminent researchers in this field from across the world.

A final word needs to be said regarding IntechOpen in particular and other publishers of open access books, which have made knowledge and research freely available to all who seek it. The cooperation and support of the IntechOpen staff in the preparation and final publication of this book is duly acknowledged.

Hassan Al-Haj Ibrahim
Professor of Chemical Engineering,
Director of Quality Assurance,
Arab University for Science and Technology,
Hamah, Syria

Chapter 1

Introductory Chapter: Pyrolysis

Hassan Al-Haj Ibrahim

1. Introduction

Pyrolysis, or thermolysis, is in essence an irreversible thermochemical treatment process of complex solid or fluid chemical substances at elevated temperatures in an inert or oxygen-free atmosphere, where the rate of pyrolysis is temperature-dependent and it increases with temperature. During pyrolysis the molecules are subjected to very high temperatures leading to very high molecular vibrations at which the molecules are stretched and shaken to such an extent that they start breaking down into smaller molecules. Pyrolysis also is always the first step in other processes such as gasification and combustion where partial or total oxidation of the treated material occurs. Thermochemical treatment processes are generally classified according to their equivalence ratio (ER), which is defined as the amount of air added relative to the amount of air required for stoichiometric combustion. The equivalence ratio for pyrolysis is 0 (ER = 0), whereas the equivalence ratio for combustion is equal to or greater than 1 ($ER \geq 1$), and the equivalence ratio for gasification varies between 0.25 and 0.50 ($Er = 0.25 - 0.50$).

The word "pyrolysis" is coined from two Ancient Greek words pyro (πυρο) meaning fire and lysis (λύσις) meaning separating (or solution), so pyrolysis means separation by fire or heat. In photolysis, by contrast, the chemical substances are treated with light rather than heat.

The simplest example of pyrolysis is food cooking. When food is cooked, the temperature of food increases leading to higher molecular vibrations and breakdown of larger complex molecules into smaller and simpler molecules which are easier to digest. Another example of pyrolysis is the pyrolysis of tobacco, paper and additives, in cigarettes and other products, which generates many volatile products including nicotine, carbon monoxide and tar that are responsible for the aroma and the hazardous health effects of smoking.

A process similar to the pyrolysis process takes place to some extent in nature, where organic substances of biological origin are buried and transformed into fossil fuels and coals of progressively higher carbon content under the action of temperature, pressure and chemical agents [1].

Pyrolysis is basically a thermal decomposition process where a raw material of high molecular weight is decomposed or cracked to produce primary volatiles. The primary thermal decomposition and dehydrogenation reactions are accompanied in general with secondary polymerization and isomerization reactions of the primary volatiles. The extent of the secondary reactions depends on the pyrolysis conditions as well as on the type of the pyrolysis reactor used. Secondary reactions are generally favored by high residence times and high temperatures. As it is practically impossible to achieve a completely oxygen-free atmosphere, there will be a small amount of oxidation reactions as well. The yields of the pyrolysis products are due to both the primary decomposition reactions of the raw material and the subsequent secondary reactions of the primary volatiles.

The end products of pyrolysis include solid residual coproducts and ash, noncondensable gases and condensable liquids known variously as pyrolysis oil, pyrolytic oil, bio-oil or tar. The type and yields of the pyrolysis products depend for the most part on the type of material treated. The pyrolysis end products can also be controlled by optimizing pyrolysis parameters such as temperature, heating rate, residence time, pressure, feed particle size and type of reactor. For example, the production of bio-oil through pyrolysis, which is a thermodynamically nonequilibrium process, requires only a short residence time in a high-temperature zone followed by rapid thermal quenching. In some pyrolysis processes, a product that is up to 80% liquid by weight can be produced.

Pyrolysis is mostly applied to organic materials. It is basically a carbonization process where an organic material of high molecular weight is decomposed or cracked to produce a solid residue with high (or higher) carbon content and some volatile products. As is well-known, any organic matter can be carbonized or made to lose progressively its atoms other than carbon to become an artificial carbon material or "carbon". In addition to organic materials, pyrolysis can be applied in certain cases to inorganic materials and to water and aqueous solutions.

Pyrolysis is an endothermic process. Determination of the overall energy balance and the thermal efficiency of the process is a fundamental step in designing an efficient pyrolysis reactor. The use of renewable energy or solar-thermal power to drive pyrolysis could make the process more economical and carbon neutral [2, 3].

Pyrolysis reactions typically occur at temperatures between 400 and 800°C. As the temperature changes, the product distribution can be altered. Lower pyrolysis temperatures usually produce more liquid and solid products, while higher temperatures favor the production of more gases as a result of more powerful thermal cracking reactions. The pyrolysis temperature has also a significant effect on the properties of the pyrolysis products. The calorific value of the pyrolytic oil, for example, increases mostly with increasing temperature.

The rate of heat transfer also influences the product distribution. In fast pyrolysis at lower temperatures, higher heating rates and small residence times favor liquid yield as the cracking of larger molecules to produce gaseous products is hindered. Liquid yield is also favored by immediate and rapid quenching which is often used to maximize the production of liquid products by condensing the vapors and gaseous molecules. Intermediate pyrolysis in screw reactors with longer residence time (minutes vs. seconds) can also be used for bio-oil production. In this process, two condensates are usually obtained, an aqueous phase and an organic phase defined as bio-oil. Although the yield of bio-oil is lower compared to fast pyrolysis, the bio-oils produced from intermediate pyrolysis are more stable, contain less oxygen, and have lower molecular weight substances, and the process is easier to control [4]. Slow pyrolysis, on the other hand, can be used to maximize the yield of solid char. This process requires slow pyrolytic decomposition at low temperatures.

Pyrolysis may be carried out at atmospheric or higher pressure or in vacuum where uncontrolled combustion is avoided. In practice, however, pyrolysis is mostly carried out at atmospheric pressure as creating a vacuum or high pressure drastically increases the cost of process equipment. Operation under high pressures results generally in greater yields of biochar and gases, while lower pressure or vacuum results in increased production of liquid products.

While feed particle size may not greatly affect pyrolysis product distribution, larger particle size tends in general to increase the liquid yield at a higher temperature range. Smaller particle sizes on the other hand favor the internal heat transfer within the particles. In fluidized bed reactors, the particles must be greater than a minimum, in order to avoid entrainment of fines, particularly where the material has a low density [5].

Reactor type is crucial for the efficient production of pyrolysis oil. Reactor types include packed or fixed-bed reactors, rotary kiln reactors and fluidized bed reactors. Fluidized bed reactors in particular, such as the auger, bubbling fluidized bed reactor and the circulating fluidized bed reactor, are highly efficient for the large commercial-scale production of pyrolysis oil [6].

Fluidized and fixed-bed reactors are mostly exterior-heating pyrolysis reactors where heat is transmitted from the exterior surface to the interior of the material. In fluidized bed reactors, fluidization increases mixing and interaction leading to efficient heat transfer, uniform temperatures, improved reaction rates and higher yield of bio-oil. On the other hand, interior heating is utilized in microwave-assisted pyrolysis with high energy effectiveness and the production of uniform products. Unlike traditional heating, microwave heating provides quick quenching of the pyrolysis vapors, which avoids secondary decomposition reactions of the primary products.

A catalyst may be used in catalytic pyrolysis processes to improve the yield and lower the temperature and/or time of reaction. Aromatic hydrocarbons such as benzene, toluene and xylenes may be directly produced by catalytic pyrolysis of biomass [7, 8].

A pyrolysis-based process has several advantages over other treatment processes:

1. The technology is relatively simple and can be made compact and lightweight. Applications of pyrolysis processes range from large-scale industrial applications where high temperatures are used to smaller-scale operations, even portable biomass conversion units, where the temperatures may be much lower. Mobile pyrolysis units for the production of liquid and solid fuels have been designed for the treatment of timber and lumber mill and other agricultural wastes. The pyrolysis units are built on trailers and consist of four basic groups: feed preparation machinery, a fluidized bed pyrolysis vessel, product separation equipment and an onboard gas turbine electrical generation system [9].

2. Pyrolysis, furthermore, can be conducted as a batch, low-pressure process, with minimal requirements for feedstock preprocessing.

3. Pyrolysis can also be used for all types of solid and liquid products and can be easily adapted to changes in feedstock composition.

4. The pyrolysis technology can be designed to produce minimal amounts of unusable byproducts.

5. In comparison with other treatment processes such as gasification, pyrolysis produces in general fewer air emissions, lower emission of nitrogen and sulfur oxides, less CO_2 generation, less dust emission and no emission of dioxin inside the pyrolyzer due to the pyrolysis with deoxidized hydrocarbon gas.

2. Industrial applications of pyrolysis

Pyrolysis is a proven and energetically efficient chemical technology that is used heavily in the chemical industry. Pyrolysis may be used in biorefineries for making a wide range of products and materials on which a future sustainable society may be based including many forms of carbon, fuels and other potentially valuable chemicals and chemical feedstocks.

There are a great many pyrolysis processes used in the production of fuels and chemicals. Such processes differ in the type of process, the use of catalysts, the substances treated and the end products. Pyrolysis processes include catalytic and noncatalytic pyrolysis, hydrous pyrolysis, vacuum pyrolysis, slow pyrolysis, torrefaction, fast pyrolysis, fluidized bed pyrolysis, flash pyrolysis, microwave-induced pyrolysis, plasma pyrolysis, empty tube pyrolysis, on-line pyrolysis and ultrasonic spray pyrolysis (USP). Other pyrolysis processes include also thermal decomposition, destructive and dry distillation, charring, tyre recycling and pyrolysis, liquefaction, high- and low-temperature carbonization, coking and thermal and catalytic cracking.

Common pyrolysis methods are frequently associated with many disadvantages including low gas yield, reducing the total energy value of the gas, and high content of tar in the gas, causing corrosion problems in the gas collection equipment and increasing the need for further treatment of the gas produced [6, 10, 11]. The disadvantages of the traditional pyrolysis methods may be overcome by radio-frequency plasma pyrolysis technology or by adding catalysts and steam.

Compared with noncatalytic pyrolysis, catalytic pyrolysis increases the pyrolytic gas and char yields but decreases the amount of oil [12]. The hydrogen concentration of the pyrolytic gas can also be considerably increased by the use of some catalysts [13]. The effect of the use of catalysts on the pyrolytic gas yield was investigated by Chen et al. It was found that some catalysts, particularly chromium oxide, have a strong positive influence on the pyrolytic gas, while other catalysts such as CuO even inhibits the pyrolytic gas yield [14]. Catalytic pyrolysis affects also the chemical composition and characteristics of the bio-oil produced. With catalytic biomass pyrolysis, the need for costly condensation and re-evaporation procedures prior to bio-oil upgrading is essentially eliminated [15, 16]. The effect of the catalysts on the yields and structure of products, however, becomes less significant with increasing temperature [17, 18].

Hydrogen may also be used in the pyrolysis process to enhance the chemical reduction and suppress oxidation by the elemental oxygen in the feedstock. The use of hydrogen can also change the pyrolysis products distribution.

Different catalysts may be used in different catalytic processes including Pt–Rh alloy, nickel-based catalysts, chromium oxide, $Co/Mo/Al_2O_3$, solid phosphoric acid and zeolite [19]. In a study conducted on biomass pyrolysis in a fixed-bed reactor, chromium oxide was used leading to gas yield improvement [14]. Oxygenated products can be reduced by utilizing zeolite-type catalysts [20]. Because of their high surface area and regular pore structures, mesoporous zeolites tend to inhibit repolymerization reactions [21, 22]. Zeolite catalyst was used in a catalytic pyrolysis process for the production of bio-oil from rice straw in a fluidized bed reactor. The water content in the bio-oil increased due to deoxygenation, and the aromatic compounds and the calorific value were also increased [12].

In hydrous or steam pyrolysis, organic materials are decomposed in the presence of superheated water or steam. The use of water as a pyrolyzing media also allows the feedstock to be introduced into the reactor in an aqueous form. The use of steam allows pyrolysis to occur at lower temperatures and higher pressures. In general, hydrous pyrolysis gives cleaner carbon with better properties and a relatively high surface area and porosity that are similar in nature to activated charcoal. The oil produced, however, contains high sulfur content and should normally be desulphurized. The C/H ratio of the pyrolytic oil is somewhat higher than that found for petroleum-derived fuels. This ratio indicates that such oil is a mixture of aliphatic and aromatic compounds. There is evidence to indicate that increasing the steam ratio (kg of steam/kg of biomass) leads to an almost linear increase of the calorific value of the biogas and an equally linear decrease in the calorific value of the biochar.

Bio-oils and fuels can be produced by hydrous pyrolysis of rice straw and other biomass materials. Steam cracking of petroleum oils can be used for the production of different cracking chemicals such as ethylene, which is a compound used for the production of many polymers and antifreeze (ethylene glycol).

According to Tu et al., radio-frequency plasma pyrolysis technology can overcome the disadvantages of common pyrolysis methods [23]. This is a capacitive dielectric heating method which employs an alternating current with high frequency and voltage to build up an electromagnetic field that produces plasma to induce the target material resulting in vigorous colliding, rubbing and thus self-heating. As the material is heated under a suitable degree of vacuum, pyrolysis occurs. The many advantages of this method include high heating rate, short heating time to reach the setting temperature, low heat loss, high concentration of syngas and low residual tar [5, 9, 24–27]. The high heating rate can efficiently decompose the combustible solid to gas products of H_2, CO, CH_4 and low carbon hydrocarbons such as C_2–C_5 [23]. The low concentration of tar in the gas phase, mostly below 10 mg/Nm^3, can be achieved because high-energy species, such as electrons, ions, atoms and free radicals, produced from the radio-frequency plasma can enhance the decomposition of tar [27].

In vacuum pyrolysis, the organic material is heated in vacuum to reduce its boiling point and also to avoid adverse chemical reactions.

In slow, or conventional, pyrolysis, the feedstock is heated slowly at a low heating rate (0.1 to 2°C per second) to low temperatures (<400°C) for a long period of time. During slow pyrolysis of biomass, the biomass is slowly devolatilized leading to the production of tar and char as the main products. The gas produced consists mainly of methane along with minor amounts of hydrogen, propane, ethylene, CO and CO_2.

Torrefaction, also known as mild pyrolysis, is an example of a slow pyrolysis process. Torrefaction of biomass is a mild form of pyrolysis carried out under atmospheric conditions and at temperatures typically ranging between 200 and 320°C, where the onset of primary pyrolysis occurs at 200°C. For the low temperatures applied in torrefaction, the warm-up period is relatively short, even for the low heating rates commonly applied in torrefaction.

Torrefaction serves to improve the properties of biomass in relation to thermochemical processing techniques for energy generation such as combustion, co-combustion with coal or gasification. Torrefaction also eliminates all biological activity reducing the risk of fire and stopping biological decomposition. About 10% of the energy content in the biomass is lost as a result of the torrefaction process, but this energy of the volatiles can be used as a heating fuel for the process itself. During torrefaction, moisture and low-weight organic volatile components are removed, and the biomass loses typically 20% of its mass (dry bone basis). In addition, torrefaction partly depolymerizes the biopolymers (cellulose, hemicellulose and lignin) and the long polysaccharide chains, producing a hydrophobic, dry, blackened solid product as "torrefied biomass" or "bio-coal" with an increased energy density (on a mass basis) and greatly increased grindability. As a result, significantly lower energy is required to process the torrefied fuel, and it no longer requires separate handling facilities when co-fired with coal in existing power stations [20]. Torrefied or so-called roasted wood has found applications as a barbecue fuel and firelighter [28]. Finally, it has been suggested that torrefied biomass is a suitable feedstock for systems previously not considered feasible for raw biomass solid fuels such as entrained flow gasification. This is because torrefied biomass forms more spherical-shaped particles during grinding or milling [29].

In fast pyrolysis, on the other hand, the organic materials are rapidly heated at 450–600°C in the absence of air in which fast heat transfer (100–1000°C/s) is applied. Achieving very high heating and heat transfer rates during pyrolysis usually requires a finely ground biomass feed. Fast pyrolysis is a well-known technique for the production of high-volatile products. Due to the short vapor residence times, products are high-quality ethylene-rich gases which can be used subsequently to produce alcohols or gasoline. The production of char and tar is considerably less in this process [30, 31].

The fast pyrolysis process has been progressively designed and optimized for producing bio-oils from biomass. A number of essential features are required for the production of bio-oil by fast pyrolysis. These include very high heating rates (1000°C/s), high heat transfer rates (600–1000 W/cm^2), short vapor residence times (typically less than 2 seconds), lower process temperatures and efficient and rapid quenching of the condensable vapors in order to prevent their cracking and hence maximize oil production [6, 32]. In experiments conducted by Lee et al., the optimum reaction temperature range for the production of bio-oil by fast pyrolysis was found to be 410–510°C [33]. The bio-oil produced by such a process may contain large molecules derived from lignin which adversely affects the bio-oil properties [34, 35].

Catalytic fast pyrolysis can be used to produce aromatics using a range of different lignocellulosic feedstocks. Catalytic fast pyrolysis has several advantages over other biomass conversion processes where pyrolysis reactions can occur in a single reactor using inexpensive aluminosilicate catalysts [36].

With the application of induction heating, a fast pyrolysis process was used for producing valuable products from rice straw, sugarcane bagasse and coconut shell in an externally heated fixed-bed reactor [37]. In one process, the straw is pulverized, dried at 150°C, mixed with other raw materials, press formed at 200°C and finally carbonized at 300–350°C [38]. In another process, the biomass mixture, after pulverization and extrusion, is oven dried and carbonized at 600–800°C [39].

Infrared radiation is an efficient technique for fast heating processes since the energy from the infrared radiation is directly transferred to the process material. Infrared radiation is used as the heating source for many applications such as food processing, surface heating, solid decomposition and fast pyrolysis of oil shale [40]. In a study by Siramard et al. on the pyrolysis of shale oil in a fixed-bed reactor with infrared heating, it was found that shale oil production is affected by the direction of the infrared beam with higher yield achieved by the cross-current in comparison with the co-current heating. This is to be explained by the fact that the residence time of the volatiles was shorter in the case of the cross-current which led to the reduction of secondary cracking reactions of the volatiles. Reduced pyrolysis pressure was also found to be beneficial to the release of volatiles and the reduction of secondary cracking reactions [40].

Fluidized bed pyrolysis is carried out in a fluidized bed created by passing an upwardly moving carrier gas stream through a bed of the solid particulate substance under appropriate conditions to cause the solid/fluid mixture to behave as a fluid. The use of a carrier gas for fluidization results in a lower calorific value of the biogas produced. A fluidized bed reactor operating at atmospheric pressure at 500°C was used to produce bio-oils from wood feedstocks and rice straw [41]. A circulating fluidized bed reactor with sand used as bed material was used at a gauge pressure of about 5–15 kPa for the production of pyrolysis oil from napier grass (*Pennisetum purpureum*) with a calorific value of 19.79 MJ/kg. The maximum pyrolysis oil production was 37 wt% at 480°C of bed temperature. The oil produced is applicable to steam engines and gas turbine engines but not to diesel engines [42].

Higher efficiency is sometimes achieved by flash pyrolysis, also called anhydrous pyrolysis. In this process, the starting material is finely divided or crushed and quickly heated to between 350 and 500°C for less than 2 seconds, generally in vacuum in order to decrease the boiling point of the byproducts and avoid adverse chemical reactions. In this process, the insulating char layer that forms at the surface of the reacting particles is continuously removed. This process is used, for example, in organic synthesis.

A flash pyrolysis process was developed by Longanbach and Bauer to produce liquid fuels, chars and gases from bituminous and subbituminous coal, municipal refuse, grass straw and other biomass materials. In this process, the biomass material is heated by contact with hot recycle char and carried in gas stream through a reactor where pyrolysis occurs at very short residence times and heat-up rates [43]. A flash pyrolysis process was also developed to convert municipal, industrial and agricultural wastes into pyrolytic oil at near ambient pressure with no need for chemicals or catalysts. At the same time, inorganics were recovered [44].

Microwave heating is an electromagnetic irradiation in the range of wavelengths from 0.01 to 1 m and the equivalent frequency range of 0.3–300 GHz. Normally, the microwave reactors for chemical synthesis and all domestic microwave ovens operate at 2.45 GHz frequency, which corresponds to a wavelength of 12.25 cm. The material which absorbs microwave irradiation is known as microwave dielectrics, and microwave heating is thus referred to sometimes as dielectric heating [45, 46].

Microwave heating has been widely used in many areas of thermochemical treatment of waste materials such as biomass, waste cooking oil and scrap tyres. This is mainly due to its high heating efficiency and easy operation. Microwave heating is an inside heating process that is carried out within the heated sample as a whole. It requires in general less energy input than conventional heating and has in addition other advantages including heating uniformity and shorter heating time [47]. In microwave-induced pyrolysis, focussed heating by microwaves makes the resulting pyrolysis different from the traditional pyrolysis. Microwave-induced pyrolysis does not require in general agitation, fluidization or a high degree of grinding, and, furthermore, it can be used for the treatment of mixed feedstocks such as municipal solid wastes [48].

According to Huang et al., higher microwave power levels contribute to higher heating rates and reaction temperatures and can therefore produce a torrefied biomass with higher calorific value and lower H/C and O/C ratios [49]. The suitable microwave power levels proposed by Wang et al. are to be set between 250 and 300 W for the torrefaction of rice husk and sugarcane residues [50]. In a study by Ahmad et al., torrefied palm kernel shell had the highest calorific value at the microwave power level of 450 W. However, when the microwave power level increased from 450 to 600 W, the calorific value of the torrefied mass decreased [45].

A work by Zhu et al. has shown that microwave heating can change the supermolecular structure of lignocellulosic materials [51]. In a study by Huang et al., it was suggested that a hydrogen-rich fuel gas (51–55% H_2) can be produced from rice straw using microwave-induced pyrolysis. The major components in the gaseous product were H_2, CO_2, CO and CH_4. Alkanes, polars and low-ringed polycyclic aromatic hydrocarbons were the three primary kinds of compounds in the liquid product. From the viewpoint of energy consumption, close to 60% of the input energy could be derived and utilized as bioenergy [52].

Bio-oils of viscosities lower than the viscosity of light and heavy fuel oil and therefore easier to handle and process were obtained by microwave-assisted pyrolysis of aspen, canola and corncob feedstocks [41].

Microwave heating was also used for the treatment of waste tyres. Experiments were run in a batch laboratory scale with an oven operating at a frequency of 2.45 GHz with a variable energy output up to 6 kW. Short pyrolysis time and manageable product properties were achieved. Typical products were a solid residue containing up to 92% of carbon, a low-viscosity oil with a high calorific value and a gas containing light hydrocarbons, hydrogen and only traces of N_2 [53].

In empty tube pyrolysis, a heated alumina or nickel tube is used in which the samples are injected. This method was developed simultaneously by two groups working at the Scottish Crop Research Institute and at Indiana University.

In ultrasonic spray pyrolysis (USP), an ultrasonic nozzle is utilized for fine chemical synthesis such as the synthesis of nanoparticles, zirconia and oxides.

Liquefaction is the thermochemical conversion of an organic solid into a liquid composed of heavy molecular compounds with characteristics similar to petroleum-based liquids such as fuel oils. Liquefaction may also involve the production of a liquid from a pyrolytic gas stream.

The materials treated by pyrolysis include:

- Solid materials such as oil shale, coal, wood, woody and herbaceous biomass and organic, agricultural and municipal solid waste materials including straw, animal dung and human fecal waste, waste plastics and even waste printed circuit boards. Rice straw in particular has several characteristics that make it an attractive lignocellulosic material for bioethanol production, such as high cellulose and hemicellulose content that can be readily hydrolysed into fermentable sugars. The high ash and silica content of rice straw, however, makes the selection of an appropriate pretreatment technique a major challenge in developing an economically viable technology for bioethanol production [54].

- Liquid materials such as petroleum fractions.

3. Other applications of pyrolysis

In addition to being a production process of chemicals and fuels, pyrolysis can be used for other purposes such as carbon-14 dating, thermal decomposition, thermal cleaning and removing of contaminants and for analysis and identification purposes as well. Pyrolysis can also be used as a pretreatment for more conventional techniques, such as incineration, gasification or steam reforming. Thermal decomposition reactions are the basis of reforming processes in the oil refining industry used to improve the combustion characteristics of gasoline and increase its octane number. In the petrochemical industry, reforming is used mainly to produce aromatic compounds which are used as feedstocks. In the treatment of plastic waste, a dual process of pyrolysis followed by steam reforming of the pyrolytic products is used for the production of gaseous fuels and hydrogen. Pyrolysis may also be used to coat a preformed substrate with a layer of pyrolytic carbon. This is typically done in a fluidized bed reactor heated to 1000–2000°C. Pyrolytic carbon coatings are used in many applications, including artificial heart valves.

A common process of thermal cleaning that is of particular interest in the oil and petrochemical industry is thermal desulphurization. Thermal desulphurization is the process where the substance to be treated is heated under atmospheric pressure in an inert atmosphere to a specified temperature and then kept at that temperature for a specified period of time. Most organic sulfur compounds undergo thermal decomposition at elevated temperatures, but some sulfur compounds decompose at lower temperatures such as many mercaptans which decompose at about 600 K and some sulphides which decompose at 530–670 K [26]. This process was found to be the most practical process for the desulphurization of petcoke and can be the only one possible when other techniques prove to be difficult or inefficient as was found in at least one case with Syrian petcoke [55].

Author details

Hassan Al-Haj Ibrahim
Arab University for Science and Technology, Hamah, Syria

*Address all correspondence to: sanjim84@yahoo.com; hasahi123@hotmail.com

References

[1] Figueiredo, Mouljin, editors. Carbon and Coal Gasification. Dordrecht: Martinues Nijhoff Publishers; 1986

[2] Yacob TW et al. Pyrolysis of human feces: Gas yield analysis and kinetic modelling. Waste Management. 2018;**79**:214-222

[3] Lichty P. Rapid high temperature solar thermal biomass gasification in a prototype cavity reactor. Journal of Solar Energy Engineering. 2010;**132**:11-12

[4] Boscagli C et al. Influence of feedstock, catalyst, pyrolysis and hydrotreatment temperature on the composition of upgraded oils from intermediate pyrolysis. Biomass and Bioenergy. 2018;**116**:236-248

[5] Conesa JA et al. Pyrolysis of polyethylene in a fluidized bed reactor. Energy & Fuels. 1994;**8**:1238-1246

[6] Bridgwater AV. Renewable fuels and chemicals by thermal processing of biomass. Chemical Engineering Journal. 2003;**91**:87-102

[7] Jae J et al. Depolymerization of lignocellulosic biomass to fuel precursors: Maximizing carbon efficiency by combining hydrolysis with pyrolysis. Energy & Environmental Science. 2010;**3**:358-365

[8] Thring RW et al. The production of gasoline range hydrocarbons from Alcell lignin using HZSM-5 catalyst. Fuel Processing Technology. 2000;**62**:17-30

[9] Wiens J. Mobile pyrolysis system for on-site biomass conversion to liquid and solid fuels. Symposium Papers: Energy from Biomass and Wastes. 1980;**4**:713-720

[10] Chen G et al. Biomass pyrolysis/gasification for product gas production: The overall investigation of parametric effects. Energy Conversion and Management. 2003;**44**:1875-1884

[11] Caldeira MIK et al. Advanced technology paths to global climate stability: Energy for a greenhouse planet. Science. 2002;**298**:981-987

[12] Choi JC et al. Bio-oil production from rice straw by the catalytic pyrolysis over zeolites. Hwahak Konghak. 2006;**44**(4):382-386

[13] Chen G et al. Catalytic pyrolysis of biomass for hydrogen rich fuel gas production. Energy Conversion and Management. 2003;**44**(14):2289-2296

[14] Chen G et al. Catalytic application to biomass pyrolysis in a fixed bed reactor. Energy Sources. 2003;**25**(3):223-228

[15] Lu Q et al. On-line catalytic upgrading of biomass fast pyrolysis products. Chinese Science Bulletin. 2009;**54**:1941-1948

[16] Samolada MC. Catalyst evaluation for catalytic biomass pyrolysis. Energy & Fuels. 2000;**14**:1161-1167

[17] Panda AK et al. Thermolysis of waste plastics to liquid fuel, a suitable method for plastic waste management and manufacture of value added products—A world prospective. Renewable and Sustainable Energy Reviews. 2010;**14**:233-248

[18] Miskolczi N et al. Thermal and thermo-catalytic degradation of high-density polyethylene waste. Journal of Analytical and Applied Pyrolysis. 2004;**72**(2):235-242

[19] Morris M. Production of bio-oils via catalytic pyrolysis. In: Handbook of Biofuels Production. Elsevier: Woodhead Publishing Series in Energy; 2011. pp. 349-389

[20] Yaman S. Pyrolysis of biomass to produce fuels and chemical feed stocks. Energy Conversion and Management. 2004;**45**:651-671

[21] Zhu Y et al. Monodispersed mesoporous SBA-15 with novel morphologies: Controllable synthesis and morphology dependence of humidity sensing. CrystEngComm. 2011;**13**:402-405

[22] Zhu Y et al. Amine-functionalized SBA-15 with uniform morphology and well-defined mesostructure for highly sensitive chemosensors to detect formaldehyde vapor. Langmuir. 2012;**28**:7843-7850

[23] Tu W et al. Products and bioenergy from the pyrolysis of rice straw via radio frequency plasma and its kinetics. Bioresource Technology. 2009;**100**:2052-2061

[24] Zhao ZL et al. Biomass pyrolysis in an argon/hydrogen plasma reactor. Chemical Engineering and Technology. 2001;**24**:197-199

[25] Merida W et al. Enhanced hydrogen production from indirectly heated, gasified biomass, and removal of carbon gas emissions using a novel biological gas reformer. International Journal of Hydrogen Energy. 2004;**29**:283-290

[26] Shie JL et al. Major products obtained from plasma torch pyrolysis of sunflower-oil cake. Energy and Fuels. 2008;**22**:75-82

[27] Hlina M et al. Plasma gasification of wood and production of gas with low content of tar. Czechoslovak Journal of Physics. 2006;**56**:B1179-B1184

[28] Girard P, Shah N. Recent developments on torrefied wood, an alternative to charcoal for reducing deforestation. REUR Technical series. 1991;**20**:101-114

[29] Bridgeman TG et al. Torrefaction of Reed Canary Grass, Wheat Straw and Willow to enhance solid fuel qualities and combustion properties. Fuel. 2008;**87**(6): 844-856. Available at: www.bioenergysite.com

[30] Meyers RA. Coal desulphurization. New York: Marcel Dekker; 1977

[31] Gibbins-Matham J, Kandiyoti R. Coal pyrolysis yields from fast and slow heating in a wire-mesh apparatus with a gas sweep. Energy & Fuels. 1988;**2**:505-511

[32] Bayerbach R, Meier D. Journal of Analytical and Applied Pyrolysis. 2009;**85**:98-107

[33] Lee K et al. Influence of reaction temperature, pretreatment, and a char removal system on the production of bio-oil from Rice straw by fast pyrolysis, using a fluidized bed. Energy & Fuels. 2005;**19**(5):2179-2184

[34] Lédé J et al. Properties of bio-oils produced by biomass fast pyrolysis in a cyclone reactor. Fuel. 2007;**86**(11-12):1800-1810

[35] Oasmaa A, Kuoppala E. Fast pyrolysis of forestry residue. 3. Storage stability of liquid fuel. Energy Fuel. 2003;**17**:1075-1084

[36] Fosteret AJ et al. optimizing the aromatic yield and distribution from catalytic fast pyrolysis of biomass over ZSM-5. Applied Catalysis A: General. 2012;**423**:154-161

[37] Tsai WT et al. Fast pyrolysis of rice straw, sugarcane bagasse and coconut shell in an induction-heating reactor. Journal of Analytical and Applied Pyrolysis. 2006;**76**(1-2):230-237

[38] Zhang L, Wang G. Method for preparation of solid fuel briquets from straw. China Patent No. CN 1699525; 2004

[39] Shen X. Method for manufacturing solid fuel from cattle manure, straw, and sawdust. China Patent No. CN 101629114; 2010

[40] Siramard S et al. Secondary cracking of volatile and its avoidance in infrared-heating pyrolysis reactor. Carbon Resources Conversion. 2018;**1**(3):202-208

[41] Luo Z et al. Research on biomass fast pyrolysis for liquid fuel. Biomass and Bioenergy. 2004;**26**(5):455-462

[42] Suntivarakorn R et al. Fast pyrolysis from Napier grass for pyrolysis oil production by using circulating fluidized bed reactor: Improvement of pyrolysis system and production cost. Energy Reports. 2018;**4**:565-575

[43] Longanbach JR, Bauer F. Fuels and Chemicals by Pyrolysis, ACS Symposium Series, 1976, 32 (Ind. Lab. Pyrolyses, Symp.)1975. pp. 476-491

[44] Pober K, Bauer H. From garbage to oil. ChemTech. 1977;**7**(3):164-169

[45] Zhu SD et al. Microwave-assisted alkali pre-treatment of wheat straw and its enzymatic hydrolysis. Biosystems Engineering. 2006;**94**(3):437-442

[46] Motasemi F, Afzal MT. A review on the microwave-assisted pyrolysis technique. Renewable and Sustainable Energy Reviews. 2013;**28**:317-330

[47] Cheng J et al. Improvement of coal water slurry property through coal physicochemical modifications by microwave irradiation and thermal heat. Energy & Fuels. 2008;**22**(4):2422-2428

[48] Karunanithy C, Muthukumarappan K. Rheological characterization of bio-oils from pilot scale microwave assisted pyrolysis, Ch. 13. In: Bernardes M, editor. Biofuel's Engineering Process Technology. Intech; 2011. pp. 293-316

[49] Huang YF et al. Microwave torrefaction of rice straw and pennisetum. Bioresource Technology. 2012;**123**:1-7

[50] Wang MJ et al. Microwave-induced torrefaction of rice husk and sugarcane residues. Energy. 2012;**37**(1):177-184

[51] Zhu SD et al. Microwave-assisted alkali pre-treatment of wheat straw and its enzymatic hydrolysis. Process Biochemistry. 2005;**40**(9):3082-3086

[52] Al-Haj Ibrahim H. Bio-energy production from rice straw. Recent Advances in Petrochemical Science. 2018;**5**(5):1-6

[53] Undri A. Microwave pyrolysis of polymeric materials: Waste tires treatment and characterization of the value-added products. Journal of Analytical and Applied Pyrolysis. 2013;**103**:149-158

[54] Al-Haj Ibrahim H. Pretreatment of straw for bioethanol production. Energy Procedia. 2012;**14**:542-551

[55] Al-Haj Ibrahim H, Ali MM. Thermal desulphurization of Syrian petroleum coke. Journal of King Saud University. 2005;**17**(2):199-212

Chapter 2

A Study on Pyrolysis of Lignin over Mesoporous Materials

Abdelrahman Mohamed Rabie and
Marwa Mohamed Abouelela

Abstract

The aromatics have widespread uses across the chemical industries. Where, the monocyclic aromatics (e.g. BTX) and phenolics compounds are important basic raw materials for several industrial petrochemical processes such as synthetic polymers, detergents, biocides, resins, explosives, etc. Traditional production of these valuable chemicals has been dependent on fossil resources for more than half a century. So, it requires strategies for alternative chemical production from renewable sources especially from nonedible biomass. This chapter presents a review of the recent literature on the fast pyrolysis process for the production of aromatic hydrocarbons using mesoporous catalysts. We focus on the factors that can enhance the yield of aromatics and the lifetime of the catalyst used. Background information on catalyst deactivation during the pyrolysis process was described. The role of mesoporous catalyst's acidity and textural and topological properties of lignin to aromatics conversion was also discussed in detail.

Keywords: pyrolysis, fast pyrolysis, lignin, mesoporous materials

1. Fast pyrolysis of lignin for production of aromatic hydrocarbons by mesoporous catalysts

Catalytic fast pyrolysis of lignin with catalysts to produce aromatic hydrocarbons has attracted many research interests in recent years. Aromatic hydrocarbons, especially benzene, toluene, ethyl benzene and xylenes (BTEX), are considered to be important and valuable chemicals in the petroleum industry (**Figure 1**) [1–5].

Pyrolysis of lignin is the thermal depolymerization of organic materials in an oxygen-free environment in a temperature range of 300–900°C [6–8]. During the thermal decomposition process, hemicellulose, cellulose, and lignin undergo different reactions, leading to a three-stage reaction: moisture removal, main depolymerization, and biochar formation [9]. Fast pyrolysis is considered one of the most economical and highly efficient technologies to convert biomass and lignin to bio-oil and high valuable chemical products [10].

Catalytic fast pyrolysis (CFP) is a further modification of fast pyrolysis directed toward the production of hydrocarbon fuels. By pyrolyzing biomass in the presence of a catalyst, it is possible to catalyze the direct production of aromatic hydrocarbons such as benzene, toluene, and xylenes [1, 3, 11–16].

Catalytic fast pyrolysis has several advantages over other biomass conversion approaches. (1) All of the desired chemistry can occur in a single reactor using

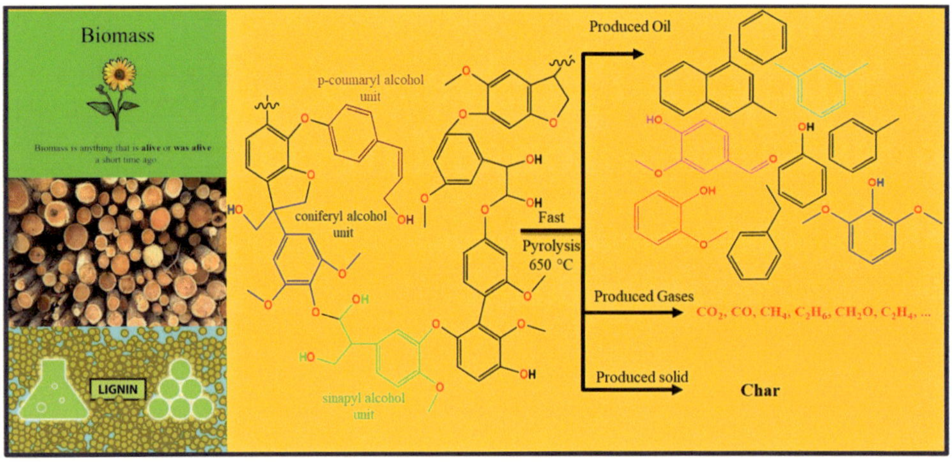

Figure 1.
Various products of fast pyrolysis of lignin.

inexpensive aluminosilicate catalysts. (2) Catalytic fast pyrolysis can be used to process a range of different lignocellulosic feedstocks with only simple pretreatment (drying and grinding, for example) prior to reaction. (3) The aromatics produced through catalytic fast pyrolysis can readily be blended into the existing gasoline infrastructure to reduce the use of crude oil [17].

Lignin is an aromatic and optically inactive amorphous heteropolymer, which is often synthesized by free radical assisted peroxidase mediated dehydrogenation of phenylpropanoid precursors, namely coniferyl alcohol, p-coumaryl alcohol, and sinapyl alcohol, joined altogether via non-hydrolysable linkages. The ratios of these three monolignols vary significantly among different plant species. For example, coniferyl alcohol is abundant in soft wood lignin while hard wood lignin comprises both coniferyl and sinapyl alcohols; however, grass lignin contains all three monolignols [18].

2. Factors that influence the pyrolysis process

The bio-oil yield from catalytic fast pyrolysis of lignin differs according to many parameters including kind of the catalyst, reactor types, temperature and the rate of heating, and reaction time, which can be 21 explained in following subdivisions.

2.1 Kind of the catalyst

The catalyst throughout the pyrolysis process causes cracking reactions and improves the quality of biomass pyrolysis products, relying on the operating conditions. The catalyst kind and reactor design play a vital role in the production of primary products during the pyrolysis process. Gaseous and liquid products can be produced by the catalytic cracking of the primary pyrolysis vapors.

Oxygenated products can be reduced by utilizing zeolite-type catalysts [19]. Zeolite-type catalysts have been studied in the pyrolysis process of lignin. Jackson et al. [20] studied the catalytic pyrolysis of lignin over KZSM-5, HZSM-5, solid phosphoric acid, Al-MCM-41, and Co/Mo/Al_2O_3. Reaction over HZSM-5 generated aromatic compounds (46.7% simple aromatics and 46.2% naphthalenic ring compounds), while reaction over Al-MCM-41 generated 17.3% simple aromatics, 13.5% naphthenic ring compounds, and 66.5% oxygenated aromatics. MCM-41 has its distinctive advantages. It is suitable for macromolecular catalytic reactions due to its larger pore size; it can make adsorption and separation, and it decreases the resistance of the molecular diffusion in pores.

As well, MCM-41 has a large specific surface area (up to 1000 m^2/g), which introduces enough sites on the surface for adsorption and catalytic reactions of reactive components, and it produces comparatively low char products [10] (**Figure 2**).

Mesoporous zeolites catalysts are favorable to inhibit the repolymerization reaction, this because the high surface area and regular pore structures [22, 23], which makes mesoporous supports suitable for the catalytic upgrading of lignocellulose [24]. Mullen and Boateng studied the pyrolysis process of four various lignin sources over CoO/MoO_3 and zeolite H-ZSM5 catalysts. They found that the H-ZSM5 catalyst was more active to produce aromatic hydrocarbons from lignin [25]. Ma et al. reported that H-USY has large enough pore size and the produced molecules during the fast pyrolysis process of alkaline lignin were able to penetrate the pores of the zeolite catalyst. The inside reaction could prevent the formation of char [26]. Thepparat et al. compared the influence of $NiMo/Al_2O_3$ and mesostructured silica catalysts on the depolymerization reaction of organosolv lignin extracted from woody eucalyptus. They found that MCM-41 and SBA-15 produced the lowest char yield [27].

2.2 Reactor types

Reactors, where the pyrolysis process takes place, play an essential role in the yield and composition of bio-oil produced from lignin pyrolysis; this is because it associates with the rate of the heating of the system, method of heat transfer, residence time of volatiles and conversion capacity of lignin. In the fast pyrolysis of lignin, the outside heating pyrolysis reactors and internal heating pyrolysis reactors are used [28].

2.3 Exterior heating pyrolysis reactors

Fluidized reactor and Fixed-bed reactor are mostly utilized in the outside heating pyrolysis system, in which the heat transmits from the exterior surface to the interior of the material. Fixed-bed reactor consists of a feeding unit, a gas flowing

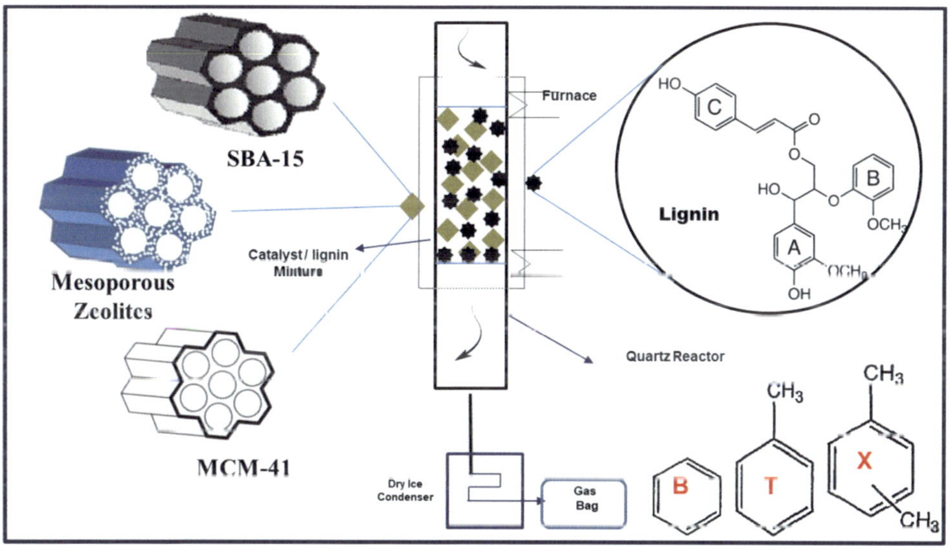

Figure 2.
Catalytic fast pyrolysis using different mesoporous materials [21].

unit, and a static bed (reaction bed), and a product exit was utilized in reactions that took place between gas phase (gaseous reactants) and solid phase (catalyst bed). Recently, Fixed-bed reactor has been widely used in the pyrolysis process of lignin or biomass. Generally, the heater that supplies the energy for pyrolysis of lignin is the electric furnace [20, 29]. The rate of heating is uneven and the efficiency of heating is low; this is because the furnace is usually attached on the periphery of the fixed bed. So, the yield of bio-oil is low and the end products are less homogenous.

However, fixed bed is appropriate for the catalytic pyrolysis process. Catalytic pyrolysis of lignin can produce High yield of aromatic compounds [30]. This is because the primary products from catalytic pyrolysis of lignin necessarily pass through the long catalytic bed (static bed), introducing more contacting opportunity for pyrolysis vapors and catalyst. Fluidized-bed reactor supplies more enormous heating for the pyrolysis process compared to fixed-bed reactor. Fluidized-bed reactor is a chamber that facilitates the mixing of gas substrates or liquid with solid particles acting as a fluid. The fluidization increases the mixing and interaction among gas and particles, resulting in efficient heat transfer symmetric temperature, and superb reaction rate.

The pyrolysis process of lignin in fluidized-bed reactor produced higher yield of bio-oil. In addition, the pyrolysis process of lignin in fluidized bed reactor produces lower guaiacols compared to the pyrolysis process in fixed-bed reactor. Even though the selectivity relies on diverse factors such as reaction conditions and the source of lignin, the high heat transfer effectiveness of fluidized-bed also secures equal high temperature during the pyrolysis process, which is contributory to demethoxylation reaction of lignin [31]. Anyhow, fluidized-bed reactor has some flaws for the pyrolysis process of lignin. The highest temperature was noticed nearly in the upper border of the bed [32] because of the continuous reaction of pyrolysis vapors in the part, indicating that undesirable side reaction happens out of the fluidized bed and in completed pyrolysis reaction in the cell of the fluidized bed, which could produce more of undesirable by-products. So, it was noticed that the most desired products were commonly gotten at the half length of the reactor [33].

2.4 Interior heating pyrolysis reactors

In fluidized bed and fixed reactors, furnace, catalyst, and hot gases are utilized as heating media, which provide the energy to the substance from the exterior surface to the interior center. These ineffectual traditional pyrolysis processes are usually producing low yield of bio-oil with low quality. Nowadays, microwave technology introduces unprecedented techniques for pyrolysis. Microwave radiation produces energy as a result of the interaction among polar molecules present in the material and electromagnetic field; the heat transmits from the interior to the exterior surface of the substance [34]. Microwave-assisted pyrolysis indicates high energy effectiveness and supplies identical interior heating. Pyrolysis process by microwave produces uniform products due to the distinctive thermal gradients [35].

Unlike traditional heating, microwave heating provides quick quenching of pyrolysis vapors, which avoid the second decomposition reactions of the primary products. Compared to pyrolysis by electric heating, the pyrolysis process by microwave assists the production of phenolic compounds from lignin [36]. In addition, esters and hydrocarbons in bio-oil steeply rose compared to that gained from electric-heating pyrolysis due to the formation of hot points formed by microwave heating, which facilitates the conversion of guaiacols to gases such CO and H_2 [37].

2.5 Temperature and the rate of heating

Temperature has a crucial effect in the catalytic fast pyrolysis of lignin. To decease the secondary reactions of primary vapors which decline the yield and quality of liquid products, a quick heating and cooling of primary vapors should be used [38]. As well as, the slow heating rate produces high yield of char [39]. Temperature has a vital influence on the yield of char and its properties. High temperatures lead to lower yield of char in all of pyrolysis reactions. The first cause for this is the removing of many volatile substances from the char at high temperatures, resulting in a decline in the char yield. For instance, the char yield declines from 31 to 17% with a rise in the temperature from 638K to 879K [40]. Low temperatures lead to imperfect decomposition of biomass resulting in a high quantity of unpyrolyzed solid materials in char content. Also, temperature has an influence on the composition of the char, chars obtained at high temperatures having high carbon content [41, 42]. At the temperatures higher than 773K,The char contains more than 85wt% carbon [43]. The yield of Liquid products increases by raising the pyrolysis temperatures up to an extreme value, at 673–823 K, but it greatly relies on other operating factors. For a diverse of feedstock sorts, it has been studied that the extreme yield of pyrolysis oil is produced at temperatures of about 673–823 K with a constant decline in char yield and rise in gas yield.

2.6 Effect of the reaction time

The time when the biomass was sustained at a specific pyrolysis temperature is the reaction time. In batch systems, the reaction time must be adequate to achieve the required result in the process. Secondary reactions of primary vapors can occur under long reaction time such as carbonization, thermal cracking, and gasification that lead low bio-oil yield [44, 45]. As well as, the reaction time is essential for the reactor configuration step. Tsai et al. [45] studied the pyrolysis process of rice husk in fixed bed reactor, they observed That, the bio-oil yield rose by rising the reaction time from 1 min to 2 min. However, the yields after that were noticed to decline slightly at long holding time.

2.7 Effect of lignin source

The source of lignin and method of isolation have a great impact on the pyrolysis process and the distribution of pyrolysis products. Wang et al. [46] studied the pyrolysis process of various four sorts of lignin, Klason lignin (KL), alkali lignin (AL), milled wood lignin (MWL), and organosolv lignin (OL), separated from the same pine wood, by TG-FTIR and Py-GC/MS as well as ^{13}C NMR spectroscopy. The ^{13}C NMR indicated chemical structure of these separated lignin's is diverse. Ether bonds such as ß-O-4 have lower thermal stability and can decompose readily at low temperatures. So, lignin's containing more ether bonds could show weaker thermal stability. Alkali lignin and milled wood lignin, because they contain more ß-O-4 bonds in their structures, have a lower temperature at maximum weight loss rate (346 and 359°C, respectively) than OL (396°C) or KL (405°C). So, the pyrolysis process of AL and MWL produced more phenolic compounds at lower temperatures. The predominate products in pyrolysis of these softwood lignin's are guaiacyl-type compounds due to shortage of syringyl-type unit in the structure of softwood and the highest yield of phenol was achieved at 600°C.

Mullen and Boateng [25] studied the pyrolysis of lignin from four various sources over an acidic zeolite (HZSM-5) and a mixed metal oxide catalyst (CoO/MoO_3). Even though two catalysts formed to be efficient catalysts for deoxygenation reaction, enhancing the formation of aromatic hydrocarbons from lignin, the acidic HZSM-5 was observed to be more active than CoO/MoO_3.

3. Deactivation of the catalysts

Phenolic compounds produced from Lignin pyrolysis are substantially adsorbed on acidic sites of zeolite catalyst, and could behave as a coke precursor and produce a great amount of coke [47]. A great amount of coke produced from catalytic fast pyrolysis of lignin commonly leads to quick deactivation of zeolite catalysts, which declines the carbon conversion effectiveness [48]. Ivanov et al. studied the deactivation of H-ZSM5 zeolite by the formation of coke and the regeneration in the production of phenol from nitrous oxide oxidation of the benzene [49]. They observed that the main reason for the deactivating effect of coke is the poisoning of active sites. A large content of coke requires to be removed for the regeneration of the catalytic activity of the catalyst.

4. The role of mesoporous catalyst acidity and textural topology

There have been many recent studies on the role of mesoporous catalysts in fast pyrolysis of lignin to aromatic compounds [20, 50–56].

In our previous publication [36] we studied the production of highly selective BTX from catalytic fast pyrolysis of lignin over supported mesoporous silica. We found that the yield of BTX rose from 17.0% in the case of MCM-48 to 32.5% in the case of Al/MCM-48 (8.4%) and 49.4% in the case of Zr/MCM-48 (2.9%) due to enhancing the acidity of the catalysts. Between the studied catalysts, Zr^{4+} loaded onto MCM-48 was considered a favorable catalyst for lignin pyrolysis with high activity and selectivity to BTX yield.

A.M. Elfadly et al [57] studied the Production of aromatic hydrocarbons from catalytic pyrolysis of lignin over acid-activated bentonite clay. They found that the catalytic fast pyrolysis of lignin over HCl-activated bentonite (mainly montmorillonite) produced a diverse of aromatics such BTX, naphthalene, indenes, and alkyl benzene, with extraordinary selectivity toward O&P-xylenes. The production of o&p-xylenes was remarkably improved at a temperature of about 550–650°C. Thus, it would be favorable to carry out the catalytic pyrolysis over HCl-activated bentonite if O&P-xylenes are the required products. They concluded that the enhanced activity of HCl-activated bentonite is due to the improvement in the textural characteristics and strong Bronsted acid sites resulted from acid treatment. H_2SO_4 and H_3PO_4 treatments formed an amorphous material and caused the destruction of the bentonite crystalline structure. Thus, these substances showed inactivity in catalytic fast pyrolysis of lignin.

Victoria B. F. Custodis et al [58] studied the Catalytic Fast Pyrolysis of Lignin over High Surface Area Mesoporous Aluminosilicates, and they focused on the influence of Acidity and Porosity. The results revealed that the acid sites (mild Bronsted and stronger Lewis) are responsible for catalyzing the pyrolysis intermediates to produce lower oxygenated phenolic compounds and aromatic hydrocarbons. MCM-41 in nano size produced a high yield and selectivity of aromatic hydrocarbons. The two most important factors are diffusion, which is affected by the pore and grain size, and the active site, which may be moderately acidic by Lewis acid sites. Nanosized grains and moderate acidity are important ingredients for a perfect catalyst for catalytic fast pyrolysis of lignin. Nanosized Al- MCM-41(50) formed the highest quantity of aromatic hydrocarbons (containing naphthalenes) (peak area 80%) compared to all mesoporous catalysts.

Lee et al. [59] studied the conversion of lignin over Al-MCM-48. Al-MCM-48 enhanced the production of light phenolic compounds extremely. The yields of aromatics and hydrocarbons were also improved by catalytic upgrading. Al-MCM-48 promoted the cracking, aromatization, and deoxygenation reactions, like decarbonylation. The hydrocarbons yield increased with increasing catalyst quantity to reach C/L = 1:5. This was due to the increased opportunity of catalytic reactions to occur on the acidic sites of the catalyst, such as cracking, decarboxylation, decarbonylation, and aromatization. Most produced hydrocarbons were cyclic and aromatic compounds, such as BTX. This indicates that phenolic compounds were transformed to cyclic and aromatic hydrocarbons by catalytic deoxygenation and cracking over Al-MCM-48 catalyst. However, the total quantity of aromatics and hydrocarbons was low compared to that of phenolic compounds. This is due to the weak acid sites of Al-MCM-48. The production of aromatics and hydrocarbons is enhanced over strong Bronsted acid sites, which are not present on Al-MCM-48. Furthermore, Si-MCM-48, which doesn't contain acid sites, exhibited lower deoxygenation effectiveness than Al-MCM-48.

Yi-Xin Chen et al [60] studied the arene production by W_2C/MCM-41-catalyzed upgrading of vapors from fast pyrolysis of lignin. The experiments were carried out in a micro pyrolyzer-gas chromatography/mass spectrometer (P-GC/MS). A range of W2C/MCM-41 catalysts with various catalyst loading quantities (Si/W) was prepared, and the activity, selectivity, and the stability degree of the catalysts were studied. They found that the catalyst with Si/W = 50:1 showed the best activity and the highest arene yield. Also, the increase in the loading percent of catalyst can enhance the cracking reaction of pyrolysis vapors. The mechanism of lignin fast pyrolysis included dehydration, demethylation, and rearrangement reactions. They concluded that MCM-41 catalysts have the best activity to catalyze the production of monocyclic arenes from primary pyrolysis vapors. In addition, they showed a high stability in catalytic fast pyrolysis of lignin. So, the modified mesoporous MCM-41 catalysts with tungsten carbide are favorable catalysts in catalytic fast pyrolysis of lignin.

Lee et al. [61] studied the catalytic fast pyrolysis of lignin over mesoporous Y zeolite using Py-GC/MS. The catalytic fast pyrolysis of lignin was taking place at 500°C using pyrolysis gas chromatography/mass spectrometry. Mesoporous Y zeolite and mesoporous material, Al-MCM-41, were tested for the catalytic fast pyrolysis of lignin. The noncatalytic pyrolysis of lignin produced phenolic compounds as a main product; this is because lignin composition mainly includes phenyl propane units. Catalytic upgrading of primary pyrolysis vapors increased the yields of low-molecular-mass phenolic compounds, monocyclic aromatics, and poly aromatic hydrocarbons (PAHs). The production of monocyclic aromatics and PAHs was increased significantly when the more acidic mesoporous Y zeolite was utilized. In contrast, the yield of alkoxy phenolic compounds was greater when the low acidic Al-MCM-41 was tested. Increasing mesoporous Y/lignin ratio showed a sharp increase in the yield of monocyclic aromatic compounds and PAHs; also, the yield of light phenolics increased, but the yield of total phenolic compounds declined.

Ming-hui Fan et al [62] studied the Catalytic Depolymerization of lignin for the production of BTX. The conversion of lignin to benzene, toluene, and xylenes (BTX) was tested over the HZSM-5 and MCM-22 catalysts; the HZSM-5 catalyst indicated the largest yield of BTX. They studied various reaction conditions, involving temperature, the catalyst/lignin ratio, and the gas flow rate. The carbon yield of BTX was nearly 25.3 C-mol% in the presence of HZSM-5 catalyst at a temperature of 550°C, a flow of N_2 300 cm^3/min, and a catalyst/lignin ratio of 2:1. HZSM-5 showed the best activity due to the mild acidity; in addition, the small pore size are useful for cracking and the deoxygenation reactions of lignin. Temperature has a high impact on the product distribution. The BTX selectivity increased by raising the temperature, but high temperature may lead to the production of olefins and alkanes due to a second cracking reaction. The BTX selectivity is strongly associated with the catalyst/lignin ratio, and the optimum ratio of catalyst to lignin was nearly 2:1. More increase of the catalyst/lignin ratio will decline the BTX yield, which leads to an increase in the yield of gas products. They observed that the reaction time should be carefully controlled to obtain high-yield BTX. The production of BTX was performed through depolymerization of lignin followed by the deoxygenation reaction.

The influence of various catalysts on the yield of aromatics during the fast pyrolysis process of biomass is summarized in **Table 1**.

Based on these studies, high quality aromatic compounds will be produced by a suitable choice of catalyst and reaction conditions.

Tang S et al. [64] studied the catalytic pyrolysis of lignin over hierarchical HZSM-5 zeolites prepared by posttreatment with alkaline solutions. They observed that the alkali treatment of HZSM-5 enhanced the catalytic activity of HZSM-5 zeolite for cracking of bulky oxygenates produced from lignin pyrolysis to form aromatic hydrocarbons. The HZSM-5 zeolite treated with 0.3 mol/L NaOH was the best choice for the catalytic fast pyrolysis of lignin for the production of aromatic hydrocarbons.

Compared to the parent HZSM-5, alkali-treated HZSM-5 zeolite showed greater selectivity to aromatic hydrocarbons by the catalytic fast pyrolysis of lignin. Some of the silicon species from the zeolite structures are removed during alkali treatment, forming new intracrystalline mesopores in the HZSM-5 grains, destroying a portion of the crystalline structure, and reducing the density of strong acid sites, which are the catalytic sites responsible for the conversion of oxygenated compounds to aromatic hydrocarbons. Even though the density of strong acid sites in zeolites is declined by alkali treatment, the effectiveness of transforming of bulky oxygenates to aromatic hydrocarbons is improved.

Catalytic pyrolysis of lignin with red mud derived hierarchical porous catalyst for the production of alkyl-phenols and hydrocarbons was studied by Wang et al. [65]. They investigated the catalytic behavior of the synthetic ACRM catalyst in the thermal decomposition of lignin. The prepared ACRM catalyst had a well-structured hierarchical porosity, which could enhance the specific surface area and the dispersion of acidic sites and active metal oxides. These enhanced properties of the ACRM showed high catalytic activity during the catalytic pyrolysis of lignin vapors. The catalyst increased the production of alkyl phenols and aromatic hydrocarbons in bio-oil by about 74% at 550°C, due to promoting the dehydroxylation, demethoxylation, demethylation, and alkylation reactions [66].

Catalyst	Type of feed	Used reactor	Pyrolysis condition	Yield of aromatics (BTX) (%)	Yield of phenols (%)	Ref
Al-MCM-41(50)	Alkaline lignin	Quartz reactor with G'C/MS	C/L ratio = 4:1, temperature at 650°C in a helium atmosphere at a heating rate of 20 cm s^{-1}, and reaction time = 20 1.	80	15	[58]
Al-MSU-J (50)2	Alkaline lignin	Quartz reactor with GC/MS	C/L ratio = 4:1, temperature at 650°C in a helium atmosphere at a heating rate of 20 cm s^{-1}, and reaction time = 20 s.	53	25	[58]
Al SBA 15(50)-2	Alkaline lignin	Quartz reactor with GC/MS	C/L ratio = 4:1, temperature at 650°C in a helium atmosphere at a heating rate of 20 cm s^{-1}, and reaction time = 20 s.	57	22	[58]
AB(HCl)	Alkaline lignin	In a down-flow fixed-bed quartz reactor	At temperature 650°C, flow of N2 = 75 ml/min and C/L ratio 3:1.	18	2	[57]
Al-MCM-48	Lignin powder (kraft, alkali)	Py-GC/MS	Pyrolysis performed at 500°C, lignin/ Al-MCM-48 - 1/5.	3.82	11.17	[59]
W_2C/ MCM-41	Alkaline lignin	Micro pyrolyzer-gas chromatography/ mass spectrometer (P-GC/MS)	Reaction conditions: 750°C, 20 s, C/L - 10:1.	14.22	0.67	[60]
	Sulfur-free lignin from wheat straw	Flow system	Pyrolysis performed at T = 550°C, f(N2) = 300 cm^3/min, and a catalyst/lignin ratio of 2.	25.3	3	[62]
MCM-22	Sulfur-free lignin from wheat straw	Flow system	Pyrolysis performed at T = 550°C, f(N2) = 300 cm^3/min, and a catalyst/lignin ratio of 2:1.	19	3.5	[63]
MCM-48	Alkaline lignin	Fixed-bed quartz reactor system	Pyrolysis performed at T = 600°C, f (N2) = 25 ml/min, and a catalyst/lignin ratio of 3:1.	17	1.12	[21]
Al/MCM-48	Alkaline lignin	Fixed-bed quartz reactor system	Pyrolysis performed at T = 550°C, f (N2) = 25 ml/min, and a catalyst/lignin ratio of 3:1.	32.5	5.6	[21]
Zr/MCM-48	Alkaline lignin	Fixed-bed quartz reactor system	Pyrolysis performed at T = 550°C, f (N2) = 25 ml/min, and a catalyst/lignin ratio of 2:1.	49.4	1.09	[21]

Table 1.
Influence of various catalysts on the yield of aromatics during the fast pyrolysis process of biomass.

Author details

Abdelrahman Mohamed Rabie* and Marwa Mohamed Abouelela
Petrochemical Department, Egyptian Petroleum Research Institute, Cairo, Egypt

*Address all correspondence to: abdo3040@yahoo.com

References

[1] Carlson TR, Tompsett GA, Conner WC, Huber GW. Aromatic production from catalytic fast pyrolysis of biomass-derived feedstocks. Topics in Catalysis. 2009;**52**:241

[2] Rezaei PS, Shafaghat H, Daud WMAW. Production of green aromatics and olefins by catalytic cracking of oxygenate compounds derived from biomass pyrolysis: A review. Applied Catalysis A: General. 2014;**469**:490-511

[3] Carlson TR, Cheng YT, Jae J, Huber GW. Production of green aromatics and olefins by catalytic fast pyrolysis of wood sawdust. Energy and Environmental Science. 2011;**4**:145-161

[4] Dhyani V, Bhaskar T. A comprehensive review on the pyrolysis of lignocellulosic biomass. Renewable Energy. 2018;**129**:695-716

[5] Wang B, Xu F, Zong P, Zhang J, Tian Y, Qiao Y. Effects of heating rate on fast pyrolysis behavior and product distribution of Jerusalem artichoke stalk by using TG-FTIR and Py-GC/MS. Renewable Energy. 2019;**132**:486-496

[6] Han X, Guo Y, Liu X, Xia Q , Wang Y. Catalytic conversion of lignocellulosic biomass into hydrocarbons: A mini review. Catalysis Today. 2019;**319**:2-13

[7] Dai L, Wang Y, Liu Y, Ruan R, He C, Duan D, et al. Bridging the relationship between hydrothermal pretreatment and co-pyrolysis: Effect of hydrothermal pretreatment on aromatic production. Energy Conversion and Management. 2019;**180**:36-43

[8] He Y, Bie Y, Lehtonen J, Liu R, Cai J. Hydrodeoxygenation of guaiacol as a model compound of lignin-derived pyrolysis bio-oil over zirconia-supported Rh catalyst: Process optimization and reaction kinetics. Fuel. 2019;**239**:1015-1027

[9] Cha JS, Park SH, Jung SC, Ryu C, Jeon JK, Shin MC, et al. Production and utilization of biochar: A review. Journal of Industrial and Engineering Chemistry. 2016;**40**:1-15

[10] Chi Y, Xue J, Zhuo J, Zhang D, Liu M, Yao Q. Catalytic co-pyrolysis of cellulose and polypropylene over all-silica mesoporous catalyst MCM-41 and Al-MCM-41. Science of the Total Environment. 2018;**633**:1105-1113

[11] Carlson TR, Jae J, Huber GW. Mechanistic insights from isotopic studies of glucose conversion to aromatics over ZSM-5. ChemCatChem. 2009;**1**:107-110

[12] Carlson TR, Jae J, Lin YC, Tompsett GA, Huber GW. Catalytic fast pyrolysis of glucose with HZSM-5: The combined homogeneous and heterogeneous reactions. Journal of Catalysis. 2010;**270**:110-124

[13] Carlson TR, Vispute TP, Huber GW. Green gasoline by catalytic fast pyrolysis of solid biomass derived compounds. Chemistry and Sustainability Energy and Materials. 2008;**1**:397-400

[14] Cheng YT, Huber GW. Chemistry of furan conversion into aromatics and olefins over HZSM-5: A model biomass conversion reaction. ACS Catalysis. 2011;**1**:611-628

[15] Jae J, Tompsett GA, Foster AJ, Hammond KD, Auerbach SM, Lobo RF, et al. Investigation into the shape selectivity of zeolite catalysts for biomass conversion. Journal of Catalysis. 2011;**279**:257-268

[16] Jae J, Tompsett GA, Lin YC, Carlson TR, Shen J, Zhang T, et al. Depolymerization of lignocellulosic biomass to fuel precursors: Maximizing carbon efficiency by combining

hydrolysis with pyrolysis. Energy and Environmental Science. 2010;**3**:358-365

[17] Foster AJ, Jae J, Cheng YT, Huber GW, Lobo RF. Optimizing the aromatic yield and distribution from catalytic fast pyrolysis of biomass over ZSM-5. Applied Catalysis A: General. 2012;**423**:154-161

[18] Duval A, Lawoko M. A review on lignin-based polymeric, micro-and nano-structured materials. Reactive and Functional Polymers. 2014;**85**:78-96

[19] Yaman S. Pyrolysis of biomass to produce fuels and chemical feedstocks. Energy Conversion and Management. 2004;**45**:651-671

[20] Jackson MA, Compton DL, Boateng AA. Screening heterogeneous catalysts for the pyrolysis of lignin. Journal of Analytical and Applied Pyrolysis. 2009;**85**:226-230

[21] Elfadly A, Zeid I, Yehia F, Rabie A, Park S-E. Highly selective BTX from catalytic fast pyrolysis of lignin over supported mesoporous silica. International Journal of Biological Macromolecules. 2016;**91**:278-293

[22] Zhu Y, Li H, Xu J, Yuan H, Wang J, Li X. Monodispersed mesoporous SBA-15 with novel morphologies: Controllable synthesis and morphology dependence of humidity sensing. CrystEngComm. 2011;**13**:402-405

[23] Zhu Y, Li H, Zheng Q , Xu J, Li X. Amine-functionalized SBA-15 with uniform morphology and well-defined mesostructure for highly sensitive chemosensors to detect formaldehyde vapor. Langmuir. 2012;**28**:7843-7850

[24] Gao D, Duan A, Zhang X, Zhao Z, E H, Li J, et al. Synthesis of NiMo catalysts supported on mesoporous Al-SBA-15 with different morphologies and their catalytic performance of DBT HDS. Applied Catalysis B: Environmental. 2015;**165**:269-284

[25] Mullen CA, Boateng AA. Catalytic pyrolysis-GC/MS of lignin from several sources. Fuel Processing Technology. 2010;**91**:1446-1458

[26] Ma Z, Troussard E, van Bokhoven JA. Controlling the selectivity to chemicals from lignin via catalytic fast pyrolysis. Applied Catalysis A: General. 2012;**423**:130-136

[27] Klamrassamee T, Laosiripojana N, Cronin D, Moghaddam L, Zhang Z, Doherty WO. Effects of mesostructured silica catalysts on the depolymerization of organosolv lignin fractionated from woody eucalyptus. Bioresource Technology. 2015;**180**:222-229

[28] Fan L, Zhang Y, Liu S, Zhou N, Chen P, Cheng Y, et al. Bio-oil from fast pyrolysis of lignin: Effects of process and upgrading parameters. Bioresource Technology. 2017;**241**:1118-1126

[29] Ferdous D, Dalai A, Bej S, Thring R, Bakhshi N. Production of H_2 and medium Btu gas via pyrolysis of lignins in a fixed-bed reactor. Fuel Processing Technology. 2001;**70**:9-26

[30] Thring RW, Katikaneni SP, Bakhshi NN. The production of gasoline range hydrocarbons from Alcell® lignin using HZSM-5 catalyst. Fuel Processing Technology. 2000;**62**:17-30

[31] Jiang G, Nowakowski DJ, Bridgwater AV. Effect of the temperature on the composition of lignin pyrolysis products. Energy & Fuels. 2010;**24**:4470-4475

[32] Kuznetsov B, Shchipko M. The conversion of wood lignin to char materials in a fluidized bed of Al• Cu• Cr oxide catalysts. Bioresource Technology. 1995;**52**:13-19

[33] Sales FG, Maranhão LC, Lima Filho NM, Abreu CA. Experimental evaluation and continuous catalytic process for fine aldehyde production from lignin. Chemical Engineering Science. 2007;**62**:5386-5391

[34] Morgan HM Jr, Bu Q, Liang J, Liu Y, Mao H, Shi A, et al. A review of catalytic microwave pyrolysis of lignocellulosic biomass for value-added fuel and chemicals. Bioresource Technology. 2017;**230**:112-121

[35] Beneroso D, Bermúdez J, Arenillas A, Menéndez J. Microwave pyrolysis of microalgae for high syngas production. Bioresource Technology. 2013;**144**:240-246

[36] Anca-Couce A. Reaction mechanisms and multi-scale modelling of lignocellulosic biomass pyrolysis. Progress in Energy and Combustion Science. 2016;**53**:41-79

[37] Bu Q, Lei H, Wang L, Wei Y, Zhu L, Zhang X, et al. Bio-based phenols and fuel production from catalytic microwave pyrolysis of lignin by activated carbons. Bioresource Technology. 2014;**162**:142-147

[38] Shafizadeh F, Chin PP. Thermal deterioration of wood. ACS Symposium Series American Chemical Society. 1977

[39] Kersten SR, Wang X, Prins W, van Swaaij WP. Biomass pyrolysis in a fluidized bed reactor. Part 1: Literature review and model simulations. Industrial and Engineering Chemistry Research. 2005;**44**:8773-8785

[40] Park WC, Atreya A, Baum HR. Experimental and theoretical investigation of heat and mass transfer processes during wood pyrolysis. Combustion and Flame. 2010;**157**:481-494

[41] Thurner F, Mann U. Kinetic investigation of wood pyrolysis. Industrial and Engineering Chemistry Process Design and Development. 1981;**20**:482-488

[42] Alves S, Figueiredo J. A model for pyrolysis of wet wood. Chemical Engineering Science. 1989;**44**:2861-2869

[43] Wang X, Kersten SR, Prins W, van Swaaij WP. Biomass pyrolysis in a fluidized bed reactor. Part 2: Experimental validation of model results. Industrial and Engineering Chemistry Research. 2005;**44**:8786-8795

[44] Bartoli M, Rosi L, Giovannelli A, Frediani P, Frediani M. Production of bio-oils and bio-char from Arundo donax through microwave assisted pyrolysis in a multimode batch reactor. Journal of Analytical and Applied Pyrolysis. 2016;**122**:479-489

[45] Tsai W, Lee M, Chang Y. Fast pyrolysis of rice husk: Product yields and compositions. Bioresource Technology. 2007;**98**:22-28

[46] Wang S, Ru B, Lin H, Sun W, Luo Z. Pyrolysis behaviors of four lignin polymers isolated from the same pine wood. Bioresource Technology. 2015;**182**:120-127

[47] Rezaei PS, Shafaghat H, Daud WMAW. Aromatic hydrocarbon production by catalytic pyrolysis of palm kernel shell waste using a bifunctional Fe/HBeta catalyst: Effect of lignin-derived phenolics on zeolite deactivation. Green Chemistry. 2016;**18**:1684-1693

[48] Bi Y, Lei X, Xu G, Chen H, Hu J. Catalytic fast pyrolysis of kraft lignin over hierarchical HZSM-5 and Hβ zeolites. Catalysts. 2018;**8**:82

[49] Ivanov D, Sobolev V, Panov G. Deactivation by coking and regeneration of zeolite catalysts for benzene-to-phenol oxidation. Applied Catalysis A: General. 2003;**241**:113-121

[50] Shen D, Zhao J, Xiao R, Gu S. Production of aromatic monomers from catalytic pyrolysis of black-liquor lignin. Journal of Analytical and Applied Pyrolysis. 2015;**111**:47-54

[51] Couhert C, Commandre JM, Salvador S. Is it possible to predict gas yields of any biomass after rapid pyrolysis at high temperature from its composition in cellulose, hemicellulose and lignin? Fuel. 2009;**88**:408-417

[52] Welker C, Balasubramanian V, Petti C, Rai K, DeBolt S, Mendu V. Engineering plant biomass lignin content and composition for biofuels and bioproducts. Energies. 2015;**8**:7654-7676

[53] Giudicianni P, Cardone G, Ragucci R. Cellulose, hemicellulose and lignin slow steam pyrolysis: Thermal decomposition of biomass components mixtures. Journal of Analytical and Applied Pyrolysis. 2013;**100**:213-222

[54] Iliopoulou E, Antonakou E, Karakoulia S, Vasalos I, Lappas A, Triantafyllidis K. Catalytic conversion of biomass pyrolysis products by mesoporous materials: Effect of steam stability and acidity of Al-MCM-41 catalysts. Chemical Engineering Journal. 2007;**134**:51-57

[55] Kelkar S, Saffron CM, Andreassi K, Li Z, Murkute A, Miller DJ, et al. A survey of catalysts for aromatics from fast pyrolysis of biomass. Applied Catalysis B: Environmental. 2015;**174**:85-95

[56] Stefanidis SD, Kalogiannis KG, Iliopoulou EF, Michailof CM, Pilavachi PA, Lappas AA. A study of lignocellulosic biomass pyrolysis via the pyrolysis of cellulose, hemicellulose and lignin. Journal of Analytical and Applied Pyrolysis. 2014;**105**:143-150

[57] Elfadly A, Zeid I, Yehia F, Abouelela M, Rabie A. Production of aromatic hydrocarbons from catalytic pyrolysis of lignin over acid-activated bentonite clay. Fuel Processing Technology. 2017;**163**:1-7

[58] Custodis VB, Karakoulia SA, Triantafyllidis KS, van Bokhoven JA. Catalytic fast pyrolysis of lignin over high-surface-area mesoporous aluminosilicates: Effect of porosity and acidity. ChemSusChem. 2016;**9**:1134-1145

[59] Lee HW, Lee IG, Park SH, Jeon JK, Suh DJ, Jung J, et al. Application of mesoporous Al-MCM-48 material to the conversion of lignin. Journal of Nanoscience and Nanotechnology. 2014;**14**:2990-2995

[60] Chen YX, Zheng Y, Li M, Zhu XF. Arene production by W2C/MCM-41-catalyzed upgrading of vapors from fast pyrolysis of lignin. Fuel Processing Technology. 2015;**134**:46-51

[61] Lee HW, Kim TH, Park SH, Jeon JK, Suh DJ, Park YK. Catalytic fast pyrolysis of lignin over mesoporous Y zeolite using Py-GC/MS. Journal of Nanoscience and Nanotechnology. 2013;**13**:2640-2646

[62] Fan MH, Deng SM, Wang TJ, Li QX. Production of BTX through catalytic depolymerization of lignin. Chinese Journal of Chemical Physics. 2014;**27**:221-226

[63] Deng SM, Fan MH, Wang TJ, Li QX. Transformation of biomass into aromatics with zeolite catalysts. Chinese Journal of Chemical Physics. 2014;**27**:361

[64] Tang S, Zhang C, Xue X, Pan Z, Wang D, Zhang R. Catalytic pyrolysis of lignin over hierarchical HZSM-5 zeolites

prepared by post-treatment with alkaline solutions. Journal of Analytical and Applied Pyrolysis. 2019;**137**:86-95

[65] Tang S, Zhang C, Xue X, Pan Z, Wang D, Zhang R. Catalytic pyrolysis of lignin over hierarchical HZSM-5 zeolites prepared by post-treatment with alkaline solutions. Journal of Analytical and Applied Pyrolysis. 2018

[66] Wang S, Li Z, Bai X, Yi W, Fu P. Catalytic pyrolysis of lignin with red mud derived hierarchical porous catalyst for alkyl-phenols and hydrocarbons production. Journal of Analytical and Applied Pyrolysis. 2018;**136**:8-17

Chapter 3

Influence of Process Parameters on Synthesis of Biochar by Pyrolysis of Biomass: An Alternative Source of Energy

Krishna Yadav and Sheeja Jagadevan

Abstract

Organic matter derived from plants and animals are known as biomass. It has a great potential to be used as an alternate source of energy by employing thermochemical conversion techniques. Among the available techniques, pyrolysis is considered to be the most efficient technique used for the conversion of biomass-based waste into value-added solid, liquid and gaseous products through heating in an oxygen-limited environment. Biochar (solid fuel) is a carbonaceous material and has multiple applications in various fields such as soil health, climate stability, water resource, energy efficiency and conservation. The yield of biochar depends on organic constituents of biomass and the pyrolytic process parameters such as temperature, time, heating rate, purging gas, particle size, catalyst, flow rate, pressure and types of pyrolysis reactors. Suitable conditions for biochar production were observed to be slow pyrolysis, low carrier gas flow rate, acid-catalysed biomass or biomass mixed with some inorganic salts, low heating rate, large particle size, high pressure, longer residence time, low temperature, feedstocks with high lignin content and pyrolysis reactors with lower bed height. Thermal conversion of biomass could be a possible sustainable alternative to provide economically viable, clean and eco-friendly solid fuel.

Keywords: biomass, biochar, process parameters, pyrolysis, solid fuel

1. Introduction

Biomass refers to organic materials derived from plants and animals and is one of the natural sources of renewable energy. Organic materials present in biomass are the most abundant bio-resource that plays a key role in carbon sequestration by capturing carbon dioxide from the atmosphere through the process of photosynthesis, thereby reducing the greenhouse gases (GHGs). Biomass has a direct effect on energy, environment and economy (3E) of any country [1]. In developing countries such as India, the contribution of biomass towards societal transformations and environment is immense as people are generally directly associated with different forms of biomass. These forms of biomass may vary from forestry, small plants, trees (woody plants), organic wastes, domestic wastes and agricultural wastes. Biomass can be employed as a renewable substitute of fossil fuels because it serves as the precursor of fossil feedstock [2]. Biomass can produce all three types of fuel, namely, solid, liquid and gaseous fuel [3]. The characteristics of biomass-derived fuel depend on several biological and thermochemical processes, and the pathway for their conversion from biomass to bioenergy is illustrated in **Figure 1**.

Biological processes such as fermentation and anaerobic digestion and thermochemical processes such as gasification, pyrolysis and liquefaction greatly influence the fuel characteristics. Out of these processes, pyrolysis is considered as the most popular and suitable method due to its lower consumption of resources and higher energy recovery [4]. Pyrolysis of biomass is a complex process in which raw biomass undergoes thermochemical conversion under oxygen-limited conditions, resulting in different products (solid, liquid and gaseous). The reaction mechanism for pyrolysis process can be summarised in three successive steps as given below [5].

Step 1: Raw biomass → moisture + unreactive residues
Step 2: Unreactive residues → $(\text{volatile} + \text{gas})_{I}$ + primary biochar
Step 3: Primary biochar → $(\text{volatile} + \text{gas})_{II}$ + secondary biochar

In the initial step, moisture is lost and the second step leads to the synthesis of primary biochar. The rate-determining final step which involves decomposition of primary biochar at very slow rate results in the production of carbon-rich solid residue, i.e., secondary biochar along with volatiles and gases. According to the International Biochar Initiative (http://www.biochar-international.org/biochar), biochar can be defined as 'a solid material obtained from the carbonisation of biomass.' The primary constituent of biochar is carbon, followed by hydrogen, oxygen, ash content and trace amounts of nitrogen and sulphur. The elemental compositions of biochar generally changes with the nature of feedstock and pyrolytic conditions such as carrier gas flow rate, catalyst, heating rate, pressure, reactor bed height, particle size, residence time and temperature [3, 6–9].

Owing to its intrinsic properties such as large surface area, porosity and surface functionality, biochar has wide application in several fields. It can be used as a precursor of activated carbon or can be used as an adsorbent for the removal of various contaminants of water and wastewater. Biochar can also be used for soil amendments and carbon sequestration, solvent recovery, vehicle exhaust emission control and separation and purification of air [1]. In addition, high heating value and low emissions make biochar as the most suitable substitute for solid fossil fuels. With advancement in technologies, the production of renewable energy has increased from 18% in 2006 to 29% in 2018 [3]. Most countries such as Australia, Finland, Germany, the United Kingdom and Turkey are reducing the generation

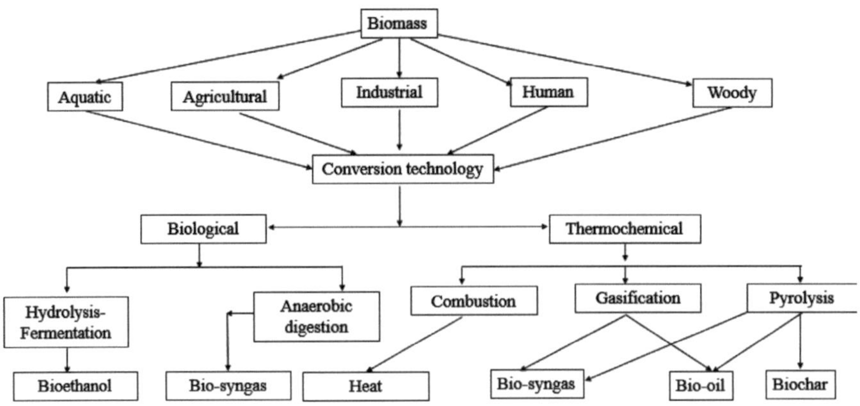

Figure 1.
Schematic pathway for conversion of biomass to bioenergy.

of electricity by thermal power plants and are encouraging generation of biomass-based electricity [10]. The primary focus of this book chapter is to delve deeper into the process parameters and know how these factors affect the characteristics of biochar produced through pyrolysis.

2. Source of biomass

Biomass can be classified based on their origin and constituents. The origin of biomass can be natural or anthropogenic (derived from processing of natural biomass), whereas the constituents of biomass may vary in terms of the polymeric structure of organic and inorganic molecules. Biomass may also be classified on the basis of their place of origin such as agricultural, animal waste, aquatic, human waste, industrial waste and woody biomass. Of these, woody biomass is abundantly found in nature, in forest area, and consists of bark, branch, chips of different trees, lumps and stem. Agricultural biomass consists of several crops, herbs and shrubs, and these biomasses ranked second in terms of abundance after woody biomass. Most countries around the world are dependent on woody and agricultural biomass for energy generation. Aquatic biomass consists of aquatic plants such as algae, blue green algae, fungi, phytoplanktons, zooplanktons and different kind of microbes. Biomass derived from humans, animal and industrial wastes includes manure from livestock; food waste generated from restaurants; waste generated from fruit industries; waste generated from paper, pulp and plastic industries and others. Energy is not only being generated and transformed into useful energy when such biomass is processed, but it also takes care of processing and handling of waste materials to a large extent.

3. Conversion techniques

It is very essential to select an appropriate technology for the extraction of energy from biomass to form a value-added product. The conversion of biomass may be achieved by two processes, namely, (i) biochemical process or (ii) thermochemical process. Biochemical and thermochemical processes make use of microorganisms and heat, respectively, to generate energy from biomass. Generation of value-added product is very less in quantity in the biochemical conversion processes. Nevertheless, this is a cost-effective and eco-friendly method and entails less production of toxic gases. Owing to the aforementioned limitations, thermochemical conversion process is a popular method and is further classified into combustion, gasification and pyrolysis.

3.1 Combustion

Combustion is considered as one of the most suitable and primitive technologies for conversion of biomass [11]. This conversion process involves a number of reactions which takes place during burning of biomass under oxygenated conditions at temperatures in the range of 800–1000°C. The biomass taken for the reaction should typically contain moisture <60% [11]. The contact between oxygen present in air and biomass determines the efficiency of the process. The efficiency can be further improved by pretreatment methods, which may enhance the total cost for the process. But the enhanced cost can be compensated by the improved process efficiency. As a result of combustion processes, various gaseous and solid products such as CO_2, H_2O, smoke, ash content and tar are produced.

3.2 Gasification

Gasification involves thermochemical conversion of carbonaceous materials into gaseous products with the help of various gaseous agents such as air, carbon dioxide, nitrogen, oxygen and mixture of gases. It is a partial oxidation occurring at elevated temperatures of 800–900°C. The primary goal of such processes is the production of gaseous products such as carbon monoxide, carbon dioxide, hydrogen and nitrogen. Solid (biochar and ash content) and liquid (bio-oil and tar) products are also produced. Biochar yield obtained during gasification is only 5–10% of raw biomass (dry basis), which is relatively lower than the yield obtained during fast pyrolysis. The composition and the heating value of the carbon and gaseous products depend upon the gaseous agents and feedstock employed. When air was employed as a gaseous medium, the syngas obtained had low heating value (4–7 MJ/Nm^3), when compared to steam as a gaseous medium (heating value of syngas varied between 10 and 15 MJ/Nm^3) [12]. The O/C ratio could be used as an indicator for determining the efficiency of the gasification process. Biomass having low O/C ratio will definitely have greater efficiency in the gasification process. In order to reduce the O/C ratio of raw feedstocks, a pretreatment step, namely, torrefaction is necessitated. Torrefaction is a thermochemical conversion process taking place at low temperature (250–280°C) and low heating rate. The product of torrefaction is brown/black colour, possessing little strength. Torrefaction enhances the energy density and hydrophobicity and reduces the weight and hydrophilic nature of biomass, thereby improving the calorific value and making the process commercially feasible for energy generation. Antal and Grønli [13] have reported that the wood yield and energy yield increased by 67–84 and ~77–90%, respectively. Similarly, retification of wood biomass also occurs at a lower range of temperature (~230–250°C), and the produced material is of reddish-brown or chocolate colour. Typically, these two processes (torrefaction and retification) demonstrated better resistance against biological attack and water loss to some extent [14].

3.3 Pyrolysis

Pyrolysis is thermal decomposition of biomass in the absence of oxygen or under oxygen-limited conditions. An oxygen-limited condition ensures that combustion of biomass does not take place. One of the primitive methods to obtain biochar is pyrolysis, which covers a wide range of thermal decomposition, thus making it very difficult to accurately define 'pyrolysis.' Based on literature survey, it could be understood that in the past scientists considered pyrolysis to be equivalent to carbonisation, wherein charcoal (solid) was the primary product. It is now well-known that pyrolysis is a process in which liquid (bio-oils) is the preferred product along with solid (char) and gas (syngas). These liquid products along with the solids depend on the operational conditions [15]. Generally, the process of pyrolysis occurs in two stages, i.e., primary stage and secondary stage. In the primary stage, dehydration, decarboxylation and dehydrogenation take place, whereas in secondary stages thermal cracking of high-molecular-weight compounds occurs, resulting in char and gaseous products such as CH_4, CO and CO_2 [3]. Based on the operational parameters, pyrolysis can be further subclassified into slow pyrolysis, fast pyrolysis, flash pyrolysis and intermediate pyrolysis.

3.3.1 Slow pyrolysis

Slow pyrolysis, also called as conventional pyrolysis, is an ancient technique employed for the conversion of biomass to charcoal. In this process, biomass is heated slowly (0.6–6.0°C/min) for longer residence times (5–30 min) at nearly

Process	Process temperature (°C)	Residence time	Solid product yield (mass %)	Carbon content of the solid product (mass %)	Carbon yield (mass carbon product/ mass carbon feedstock)	References
Slow pyrolysis	~400	Min. to days	~30	95	~0.58	[16]
Fast pyrolysis	~500–1000	~1 s	12–26	74	0.2–0.26	[13, 17]
Gasification	~800	~10–20 s	~10	—	—	[16]
Flash carbonisation	~300–600	<30 min	37	~85	~0.65	[13]
Torrefaction	~290	10–60 min	61–84	51–55	0.67–0.85	[16]

Table 1.
Comparison of typical operational conditions and product properties associated with various processes for biochar production.

500°C. During slow pyrolysis, the production of solid (biochar) is found to be maximum when compared to liquid and gaseous products. The lower heating rate and higher vapour residence times enable the process to complete the secondary reactions. However, the vapours produced during the reactions of slow pyrolysis do not escape unlike fast pyrolysis [15].

3.3.2 Fast pyrolysis

Fast pyrolysis can be fundamentally differentiated from slow pyrolysis. Fast pyrolysis requires smaller-sized biomass (1–2 mm) having moisture content <10%, increased temperature (500–1000°C) and heating in very short span of time (1–5 s). Generally, fast pyrolysis favours liquid (bio-oil) production which contributes to 60–75% as compared to char (15–25%) and noncondensable gases (10–20%). The high temperature and minimum residence time during fast pyrolysis enable thermal cracking of biomass, resulting in char formation [15]. In addition, the high heating rate induces thermal conversion of raw biomass into liquid product before the formation of biochar [3].

Owing to its wide applicability, pyrolysis is considered as one of the most suitable techniques amongst existing technologies for conversion of biomass into value-added products. The peculiarity of this technique is its inherent flexibility, and more importantly it works based on desired products. Slow pyrolysis is recommended for higher biochar yield, whereas fast pyrolysis is the preferred choice for higher yield of bio-oil (**Table 1**). The characteristics associated with products of pyrolysis could be changed by varying the operational parameters. The higher cost required for pyrolysis could be compensated by setting up larger pyrolysis plants.

4. Effects of process parameters on yield of pyrolytic products

Pyrolysis is generally performed to enhance the yield of the pyrolytic products, and hence it is very essential to discuss the effect of process parameters on the production of biochar. Pyrolytic conditions can markedly affect the quantitative characteristics of the pyrolysed products.

4.1 Effect of carrier gas flow rate

In pyrolysis, the carrier gas flow rate influences the yield of products by affecting the generation of vapours during pyrolysis. If these vapours are not purged out, then they may take part in secondary reactions which definitely affects the composition and nature of pyrolysis products. There are several purging agents available such as nitrogen, argon, steam and carbon dioxide, but nitrogen is the most popular owing to its cost-effectiveness, easy availability and inert behaviour. Increase in flow rate of nitrogen from 1.2–4.5 to 50–400 mL/min led to a decrease in the biochar yield of 24.4–22.6 and 28.48–27.21%, respectively [18].

4.2 Effect of catalysts

The distribution of pyrolytic products can be affected by the presence of catalysts. Based on the catalytic effects, catalysts may be classified as primary and secondary catalysts. Primary catalysts are generally mixed prior to pyrolysis either in the dry mode or wet mode, while secondary catalysts are placed into the secondary chamber downstream to primary chamber where pyrolysis is taking place. Several catalysts such as alumina, Al-MCM-41, oxides of magnesium, oxides of nickel and ZSM-5 showed positive effects on the yield of biochar [19]. The yield of liquid (tar) and gaseous product reduces after mixing of inorganic salt to the biomass, whereas increase in the yield of solid products was observed [20]. Additionally, the yield of biochar was found to be increased after acid and base modification of biomass. Biomass catalysed with ZSM-5 demonstrated higher biochar yield than biomass modified with alumina oxide and sodium carbonate [21]. In contrary to this, some studies showed that ZSM-5 was responsible for the reduction in biochar yield, which was in accordance with the results obtained by Zhang et al. [22].

4.3 Effect of heating rate

The rate of change of heat during pyrolysis can affect the distribution of solid, liquid and gaseous products. Secondary stages of pyrolysis and thermal cracking are lacking at lower heating rates which favours the production of char. At higher heating rate, fragmentation of biomass takes place which is responsible for a higher yield of liquid and gaseous products. During pyrolysis of feedstocks such as leaves of *Ferula orientalis*, seed of safflower and leaves of *Carthamus tinctorius*, a decrement in the yield of biochar was observed as temperature was increased from 400 to 500°C along with a concurrent increase in the heating rate from 30 to 50°C/min [23, 24]. Production of bio-oil is also found to be affected by changing the heating rate. At increased heating rate from 500 to 700°C/min, 8% increment in the production of bio-oil from sawdust was observed, while there was no change observed when heating rate was increased from 700 to 1000°C/min [25]. Similar results were also observed from the biomass of cottonseed cake when it underwent pyrolysis, and heating rate was increased from 5 to 300°C/min; an increase (9%) in production of bio-oil was observed. However no change in yield of bio-oil was observed on further increase of heating rate from 300 to 700°C/min [26]. Interestingly, a reduction in the oxygen content was observed as a result of high heating rate during pyrolysis [27].

4.4 Effect of pressure

Pyrolysis carried out under increased pressure is also known to affect the yield of biochar. When the pressure inside the reactor is higher than the atmospheric pressure, the yield of biochar gets enhanced [28]. High pressure is responsible for increasing the residence times due to which volatile matter is added to the carbonaceous material, thus resulting in high yield of biochar [29]. The volatile matter produced during pyrolysis is responsible for the generation of liquid and gaseous products as discussed in fast pyrolysis. This could be used for high biochar production either by increasing the pressure or decreasing the heating rate [29]. The concentration of carbon is also found to be enhanced by increasing pressure, which ultimately enhances the energy density of the produced biochar [13].

4.5 Effect of particle size

Biomass, being a poor conductor of heat, may create difficulties in the transfer of heat during pyrolysis. To negate this effect, particle size should be taken into consideration. As the particle size increases, the distance between the surface of biomass and the core end also increases which retards the flow of heat, resulting in high yield of solid biochar [30]. Longer-sized biomass particles were found to be responsible for reduction in the yield of liquid products due to occurrence of secondary reactions at temperatures higher than 527°C [31]. During pyrolysis, as the particle size increased from 17 to 20 mm, the water content also increased from 40 to 55% with a corresponding decrease in the carbon content of solid biochar from 78.5 to 75% [32].

The yield of biochar is also influenced by the shape of the particles. Slab-shaped and cylindrical-shaped biomass favours higher yield of biochar, whereas spherical shape favours lesser yield of biochar [33]. Some studies showed that the rate of heat and mass transfer was found to be slower with particles of spherical nature. In addition, the orientation of particles also plays an important role in explaining the pyrolytic behaviour of biomass. The permeability of flow along the particles was reported to be 10,000 times higher than that across the particles, although thermal conductivity along the particles was found to be two times higher than across the particles [34]. Heating of the perpendicular-oriented particles increased the yield of biochar and gaseous products but decreased the tar yield.

4.6 Effect of residence time

The residence time between volatile matter generated during pyrolysis and the hot biochar affects the yield of biochar. Longer residence time provides a greater chance for repolymerisation of the constituents of biomass, thereby enhancing the yield of biochar, while lower residence time decreases the yield of biochar. Popular wood and yellow brown coal biomass which underwent pyrolysis for longer residence times yielded higher biochar [35, 36]. Pyrolysis of sorghum bagasse at optimum temperatures of 525°C resulted in an increase in biochar and gaseous yield with corresponding decrease in bio-oil production (75–57%) on increasing the residence time from 0.2 to 0.9 s [37].

On contrary to the aforementioned results, some studies showed that residence time not only affects the biochar yield, but they also greatly influence the composition and quality of gaseous and liquid products. The porosity of biochar also increases due to higher residence time [38]. Fassinou et al. [39] showed an interactive effect of temperature and residence time, wherein increased temperature and residence times resulted in increase in the biochar yield, whereas lower temperature and increase in contact time reduced the yield of biochar. It is therefore difficult to make direct inferences regarding the relation between the production of biochar and the residence times.

4.7 Effect of temperature

Pyrolytic temperature is negatively correlated with the biochar yield. As the temperature increases, thermal cracking of high-molecular-weight hydrocarbons present in biomass increases which further increases the production of liquid and gaseous products, but the biochar yield decreases. Pyrolysis of biomass (hazelnut shell and sesame stalk) occurring at temperatures between 400 and 700°C resulted in decrease in yield of biochar by 10% for hazelnut shell and 17% for sesame stalk [40–42]. On increasing temperature from 365 to 606°C, a corresponding decrease in biochar yield from 31 to 17% was observed [33]. While at lower temperatures, the yield of biochar was found to be high, probably due to the partial decomposition of biomass.

The composition of biochar also gets affected by varying the temperature in which carbon content was found to be more than 85% (by dry wt.) in biochar synthesised at 500°C [32]. As temperature increases, the calorific value of produced biochar also increases [14]. The production of liquid during pyrolysis was found to be maximum as temperatures increased from 400 to 600°C. Additionally, above this range of temperature, the decomposition of vapours produced as a result of secondary reactions become more active which decreases the yield of liquid products [33]. The optimum yield of bio-oil in terms of calorific values and H/C ratio was achieved at pyrolytic temperature of 500°C [43]. However, these characteristics greatly depend on other parameters too. Pyrolysis occurring at temperatures >500°C increases the higher heating value of noncondensable gases such as carbon monoxide and methane [44].

4.8 Effect of compositions of biomass

The biomass constituents also affect the nature and composition of pyrolytic products. Several studies have reported biomass to be made up of cellulose, hemicellulose and lignin, which influence the nature of products of thermochemical processes. The production of biochar from lignin-based biomass is the result of breaking of weak bonds [45]. Some studies showed that the rate of lignin degradation depend upon the amount of lignin present in the respective biomass. The place of origin of biomass also greatly affects the production of char. Biomass obtained coniferous feedstocks are capable of producing more and stable biochar than deciduous lignin [46]. At lower temperatures, cellulose degradation leads to the formation of a more stable anhydrocellulose thereby resulting in a higher biochar yield. While at higher temperatures, cellulose is converted into more volatile compounds [47]. Furthermore, biomass rich in cellulose and hemicellulose are easily converted into a mixture of liquid products. However, lignin-rich biomass is responsible for the generation of solid product such as biochar [48]. As cellulose content in biomass increases, there is corresponding enhancement in the gaseous products, but yield of tar and char decreases. Additionally, structural differences in biomass content could lead to the compositional changes in pyrolytic products. The presence of moisture content in biomass will also affect the pyrolysis products. Studies suggested that biomass having moisture content more than 30% are not suitable for pyrolysis [1]. Biomass with low moisture content favours biochar production [49], while those loaded with high moisture content favours bio-oil production [50, 51].

4.9 Effect of reactor configuration

The thermal behaviour of various biomass has been studied with different types of pyrolysis reactors (**Figure 2**). The most exciting factor which influences the yield

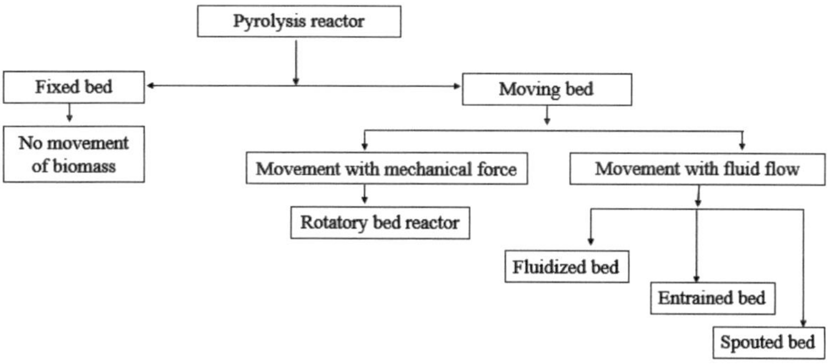

Figure 2.
Different types of pyrolysis reactor.

of pyrolytic product is the transfer of heat from the reactor to biomass particles. The transfer of heat could take place either through solid-solid mode or gas-solid mode. In the case of fixed bed reactors, solid-solid mode of heat transfer is favoured, while in the case of moving bed reactors, the dominating mode of heat transfer is a combination of conduction and convection [52].

The height of these bed reactors is also an important factor which influences the yield of pyrolytic products. The yield of biochar in both reactors remains the same, but the yield of liquid and gaseous products differs [53]. As the bed height of the reactor increases from 5 to 10 cm, reduction in the biochar yield was observed from 28.48 to 25.04% [54], while further increment in bed height results in an increase in biochar yield [22]. Higher bed height is responsible for longer vapour residence time, due to repolymerisation of biomass particles which increases the biochar yield, but an opposite effect was observed in the case of smaller bed height reactors due to less time available for repolymerisation of volatile matters.

5. Application of biochar

Biomass is a renewable substitute and precursor of fossil fuels which could be converted into different forms of valuable products. Out of these, biochar is considered as the most important product due to its intrinsic properties. Biochar has wide application in different fields such as soil, air, water and energy. In order to know the full potential of applications of biochar, it is imperative to conduct process optimization alongside minimal energy inputs and costs associated with wide applicability of the by-products at larger scales.

6. Summary and future prospects

Pyrolysis is a thermochemical conversion process in which biomass is converted into a valuable product, i.e., biochar. Synthesis of biochar depends on various process parameters such as carrier gas, catalyst, heating rate, particle size, pressure, flow rate, residence time, temperature, composition of feedstocks and types of pyrolysis reactors. Feedstocks having high cellulose content will favour tar formation, and biomass with high lignin content favours biochar formation. Suitable conditions for biochar production were observed to be slow pyrolysis, low carrier gas flow rate, acid-catalysed biomass or biomass mixed with some inorganic salts, low heating rate, large particle size, high pressure, longer residence time, low temperature, the nature of feedstocks and lower reactor bed height. Further studies need to be done in order to incorporate the interactive effects of these process parameters by employing different statistical tools.

Acknowledgements

This work is supported by FRS Scheme of IIT(ISM) (Ref No. FRS/86/2014-2015/ESE).

Conflict of interest

There are no conflicts to declare.

Author details

Krishna Yadav and Sheeja Jagadevan*
Department of Environmental Science and Engineering, Indian Institute of Technology (Indian School of Mines), Dhanbad, Jharkhand, India

*Address all correspondence to: sheejaj@iitism.ac.in

References

[1] Demirbas A. Effects of temperature and particle size on bio-char yield from pyrolysis of agricultural residues. Journal of Analytical and Applied Pyrolysis. 2004;**72**:243-248. DOI: 10.1016/j.jaap.2004.07.003

[2] Collard FX, Blin J. A review on pyrolysis of biomass constituents: Mechanisms and composition of the products obtained from the conversion of cellulose, hemicelluloses and lignin. Renewable and Sustainable Energy Reviews. 2014;**38**:594-608. DOI: 10.1016/j.rser.2014.06.013

[3] Tripathi M, Sahu JN, Ganesan P. Effect of process parameters on production of biochar from biomass waste through pyrolysis: A review. Renewable and Sustainable Energy Reviews. 2016;**55**:467-481. DOI: 10.1016/j.rser.2015.10.122

[4] Angın D. Effect of pyrolysis temperature and heating rate on biochar obtained from pyrolysis of safflower seed press cake. Bioresource Technology. 2013;**128**:593-597

[5] Demirbas A, Caglar A, Akdeniz F, Gullu D. Conversion of olive husk to liquid fuel by pyrolysis and catalytic liquefaction. Energy Sources, Part A: Recovery, Utilization, and Environmental Effects. 2010;**22**:631-639. DOI: 10.1080/00908310050045582

[6] Kambo HS, Dutta A. A comparative review of biochar and hydrochar in terms of production, physico-chemical properties and applications. Renewable and Sustainable Energy Reviews. 2015;**45**:359-378. DOI: 10.1016/j.rser.2015.01.050

[7] Bhattacharjee N, Biswas AB. Pyrolysis of *Alternanthera philoxeroides* (alligator weed): Effect of pyrolysis parameter on product yield and characterization of liquid product and bio char. Journal of the Energy Institute. 2018;**91**:605-618. DOI: 10.1016/j.joei.2017.02.011

[8] Yuan H, Lu T, Wang Y, et al. Influence of pyrolysis temperature and holding time on properties of biochar derived from medicinal herb (*Radix isatidis*) residue and its effect on soil CO_2 emission. Journal of Analytical and Applied Pyrolysis. 2014;**110**:277-284. DOI: 10.1016/j.jaap.2014.09.016

[9] Ahmad M, Rajapaksha AU, Lim JE, et al. Biochar as a sorbent for contaminant management in soil and water: A review. Chemosphere. 2014;**99**:19-23. DOI: 10.1016/j.chemosphere.2013.10.071

[10] Bilge S, Keles S, Kaygusuz A, et al. Global warming and renewable energy sources for sustainable development: A case study in Turkey. Renewable and Sustainable Energy Reviews. 2008;**12**:372-396. DOI: 10.1016/j.rser.2006.07.016

[11] Nussbaumer T. Combustion and co-combustion of biomass: Fundamentals, technologies, and primary measures for emission reduction. Energy and Fuels. 2003;**17**:1510-1521

[12] Kumar A, Jones DD, Hanna MA. Thermochemical biomass gasification: A review of the current status of the technology. Energies. 2009;**2**:556-581. DOI: 10.3390/en20300556

[13] Antal MJ, Grønli M. The art, science, and technology of charcoal production. Industrial and Engineering Chemistry Research. 2003;**42**:1619-1640

[14] Mohan D, Sarswat A, Ok YS, Pittman CU. Organic and inorganic contaminants removal from water with biochar, a renewable, low cost

and sustainable adsorbent—A critical review. Bioresource Technology. 2014;**160**:191-202. DOI: 10.1016/j.biortech.2014.01.120

[15] Mohan D, Pittman CU, Steele PH. Pyrolysis of wood/biomass for bio-oil: A critical review. Energy and Fuels. 2006;**20**:848-889. DOI: 10.1021/ef0502397

[16] Bridgwater AV. The production of biofuels and renewable chemicals by fast pyrolysis of biomass. International Journal of Global Energy Issues. 2007;**27**:160-203. DOI: 10.1504/IJGEI.2007.013654

[17] Desisto WJ, Ii NH, Beis SH, et al. Fast pyrolysis of pine sawdust in a fluidized-bed reactor. Energy and Fuels. 2010;**24**:2642-2651. DOI: 10.1021/ef901120h

[18] Ertaş M, Hakki Alma M. Pyrolysis of laurel (*Laurus nobilis* L.) extraction residues in a fixed-bed reactor: Characterization of bio-oil and bio-char. Journal of Analytical and Applied Pyrolysis. 2010;**88**:22-29. DOI: 10.1016/j.jaap.2010.02.006

[19] Stefanidis SD, Kalogiannis KG, Iliopoulou EF, et al. In-situ upgrading of biomass pyrolysis vapors: Catalyst screening on a fixed bed reactor. Bioresource Technology. 2011;**102**:8261-8267. DOI: 10.1016/j.biortech.2011.06.032

[20] Encinar JM, Beltran FJ, Ramiro A, Gonza JF. Catalyzed pyrolysis of grape and olive bagasse. Influence of catalyst type and chemical treatment. Industrial and Engineering Chemistry Research. 1997;**36**:4176-4183

[21] Smets K, Roukaerts A, Czech J, et al. Slow catalytic pyrolysis of rapeseed cake: Product yield and characterization of the pyrolysis liquid. Biomass and Bioenergy. 2013;**57**:180-190

[22] Zhang H, Xiao R, Huang H, Xiao G. Comparison of non-catalytic and catalytic fast pyrolysis of corncob in a fluidized bed reactor. Bioresource Technology. 2009;**100**:1428-1434. DOI: 10.1016/j.biortech.2008.08.031

[23] Aysu T, Küçük MM. Biomass pyrolysis in a fixed-bed reactor: Effects of pyrolysis parameters on product yields and characterization of products. Energy. 2014;**64**:1002-1025. DOI: 10.1016/j.energy.2013.11.053

[24] Angın D. Pyrolysis of safflower (*Carthamus tinctorius* L.) seed press cake: Part 1. The effects of pyrolysis parameters on the product yields. Bioresource Technology. 2008;**99**:5492-5497. DOI: 10.1016/j.biortech.2007.10.046

[25] Salehi E, Abedi J, Harding T. Bio-oil from sawdust: Pyrolysis of sawdust in a fixed-bed system. Energy and Fuels. 2009;**23**:3767-3772

[26] Pehlivan E, Özbay N, Yargıç AS, Şahin RZ. Production and characterization of chars from cherry pulp via pyrolysis. Journal of Environmental Management. 2017;**203**:1017-1025. DOI: 10.1016/j.jenvman.2017.05.002

[27] Akhtar J, Saidina N, Wood P. A review on operating parameters for optimum liquid oil yield in biomass pyrolysis. Renewable and Sustainable Energy Reviews. 2012;**16**:5101-5109. DOI: 10.1016/j.rser.2012.05.033

[28] Antal MJ, Mok WSL. Review of methods for improving the yield of charcoal from biomass. Energy and Fuels. 1990;**4**:221-225

[29] Antal MJ, Allen SG, Dai X, et al. Attainment of the theoretical yield of carbon from biomass. Industrial and Engineering Chemistry Research. 2000;**39**:4024-4031. DOI: 10.1021/ie000511u

[30] Encinar JM, Gonzalez JF, Gonzalez J. Fixed-bed pyrolysis of *Cynara cardunculus* L. product yields and compositions. Fuel Processing Technology. 2000;**68**:209-222

[31] Di Blasi C, Chimica I, Ii F. Modeling intra- and extra-particle processes of wood fast pyrolysis. AICHE Journal. 2002;**48**:2386-2397

[32] Wang X, Kersten SRA, Prins W, Van Swaaij WPM. Biomass pyrolysis in a fluidized bed reactor. Part 2: Experimental validation of model results. Industrial and Engineering Chemistry Research. 2005;**44**:8786-8795. DOI: 10.1021/ie050486y

[33] Park WC, Atreya A, Baum HR. Experimental and theoretical investigation of heat and mass transfer processes during wood pyrolysis. Combustion and Flame. 2010;**157**:481-494. DOI: 10.1016/j.combustflame.2009.10.006

[34] Roberts AF. The heat of reaction during the pyrolysis of wood. Combustion and Flame. 1971;**17**:79-86

[35] Kim KH, Eom IY, Lee SM, et al. Investigation of physicochemical properties of biooils produced from yellow poplar wood (*Liriodendron tulipifera*) at various temperatures and residence times. Journal of Analytical and Applied Pyrolysis. 2011;**92**:2-9. DOI: 10.1016/j.jaap.2011.04.002

[36] Yeasmin H, Mathews JF, Ouyang S. Rapid devolatilisation of Yallourn brown coal at high pressures and temperatures. Fuel. 1999;**78**:11-24

[37] Scott DS, Majerski P, Piskorz J, Radlein D. A second look at fast pyrolysis of biomass—The RTI process. Journal of Analytical and Applied Pyrolysis. 1999;**51**:23-37

[38] Tsai WT, Chang CY, Lee SL. Preparation and characterization of activated carbons from corn cob. Carbon. 1997;**35**:1198-1200

[39] Fassinou WF, Van De Steene L, Toure S, et al. Pyrolysis of Pinus pinaster in a two-stage gasifier: Influence of processing parameters and thermal cracking of tar. Fuel Processing Technology. 2008;**90**:75-90. DOI: 10.1016/j.fuproc.2008.07.016

[40] Putun AE, Ozcan A, Putun E. Pyrolysis of hazelnut shells in a fixed-bed tubular reactor: Yields and structural analysis of bio-oil. Journal of Analytical and Applied Pyrolysis. 1999;**52**:33-49

[41] Choi HS, Choi YS, Park HC. Fast pyrolysis characteristics of lignocellulosic biomass with varying reaction conditions. Renewable Energy. 2012;**42**:131-135. DOI: 10.1016/j.renene.2011.08.049

[42] Ates F, Putun E, Putun AE. Fast pyrolysis of sesame stalk: Yields and structural analysis of bio-oil. Journal of Analytical and Applied Pyrolysis. 2004;**71**:779-790. DOI: 10.1016/j.jaap.2003.11.001

[43] Scott DS, Piskorz J, Bergougnou MA, et al. The role of temperature in the fast pyrolysis of cellulose and wood. Industrial and Engineering Chemistry Research. 1988;**27**:8-15. DOI: 10.1021/ie00073a003

[44] Luo Z, Wang S, Liao Y, et al. Research on biomass fast pyrolysis for liquid fuel. Biomass and Bioenergy. 2004;**26**:455-462. DOI: 10.1016/j.biombioe.2003.04.001

[45] Santos RB, Hart PW, Jameel H, Chang H. Wood based lignin reactions important to the biorefinery and pulp and paper industries. BioResources. 2013;**8**:1456-1477

[46] Brebu M, Vasile C. Thermal degradation of lignin—A review.

Cellulose Chemistry and Technology. 2010;**44**:353-363

[47] Demirbas A. Bio-fuels from agricultural residues. Energy Sources, Part A: Recovery, Utilization, and Environmental Effects. 2008;**30**:101-109. DOI: 10.1080/00908310600626788

[48] Yang H, Yan R, Chen H, et al. In-depth investigation of biomass pyrolysis based on three major components: Hemicellulose, cellulose and lignin. Energy and Fuels. 2006;**20**:388-393

[49] Demirbas A. Effect of initial moisture content on the yields of oily products from pyrolysis of biomass. Journal of Analytical and Applied Pyrolysis. 2004;**71**:803-815. DOI: 10.1016/j.jaap.2003.10.008

[50] Xiong S, Zhuo J, Zhang B, Yao Q. Effect of moisture content on the characterization of products from the pyrolysis of sewage sludge. Journal of Analytical and Applied Pyrolysis. 2013;**104**:632-639. DOI: 10.1016/j.jaap.2013.05.003

[51] Huang YF, Kuan WH, Lo SL, Lin CF. Hydrogen-rich fuel gas from rice straw via microwave-induced pyrolysis. Bioresource Technology. 2010;**101**:1968-1973. DOI: 10.1016/j.biortech.2009.09.073

[52] Bridgwater AV, Meier D, Radlein D. An overview of fast pyrolysis of biomass. Organic Geochemistry. 1999;**30**:1479-1493

[53] Aylón E, Fernández-Colino A, Navarro MV, et al. Waste tire pyrolysis: Comparison between fixed bed reactor and moving bed reactor. Industrial and Engineering Chemistry Research. 2008;**47**:4029-4033

[54] Meesuk S, Cao J, Sato K, et al. The effects of temperature on product yields and composition of bio-oils in hydropyrolysis of rice husk using nickel-loaded brown coal char catalyst. Journal of Analytical and Applied Pyrolysis. 2012;**94**:238-245. DOI: 10.1016/j.jaap.2011.12.011

Chapter 4

Modeling and Optimization of Product Profiles in Biomass Pyrolysis

Udaya Bhaskar Reddy Ragula, Sriram Devanathan and Sindhu Subramanian

Abstract

Biomass feed comes in many varieties, but have common chief constituents of hemicellulose, cellulose, and lignin. As the relative proportions of these constituents may vary, customization of the pyrolysis process conditions is required to produce a desired product profile. By recognizing the sources of variation, the reactor settings may be intelligently controlled, to achieve optimal operation. These considerations include biomass classification, feed rate, moisture content, particle size, and inter-particle thermal gradients (which arise during pyrolysis based on heating rate and temperature distribution). This chapter addresses the optimization of product profiles during biomass pyrolysis from a modeling perspective. Fundamental models for packed bed and fluidized bed pyrolyzers are developed, using kinetics from existing literature. The proposed optimization approach (inclusive of the kinetic and process models) can guide practical achievement of desired product profiles of the biomass pyrolysis process.

Keywords: biomass pyrolysis, product profiles, kinetics, modeling and optimization

1. Introduction

The chief constituents of any biomass are hemicellulose, cellulose, lignin and inerts [1]. The amounts of the chief constituents of various biomasses along with the elemental compositions are presented in **Table 1**. Generally, pyrolysis is a process where biomass or other carbonaceous material is heated in the absence of air supply or in the presence of inert gas supply such as nitrogen. During the biomass pyrolysis, the bonds between high molecular compounds are broken and low molecular weight compounds are formed. The range of pyrolysis products are gases, liquids, char, and ash. The ash is the inert material present in biomass. The fractions of these products depend on composition of biomass and process conditions such as heating rate, biomass feed rate, particle size, moisture content, and the rate of heat transfer. The rate of heat transfer depends greatly on mixing conditions in the pyrolyzers.

Table 1 provides the chief constituents of biomass in wt% on dry and ash free basis along with the elemental composition for various biomasses. These chief constituents will vary depending on the type, location, and age of the plant. The error in values reported in **Table 1** is 2–5%. The wt% of hemicellulose varies from

Type of biomass	Biomass constituents			Elemental composition					Ref.#
	HEMI* (%)	CEL* (%)	LIG* (%)	C (%)	H (%)	N (%)	S (%)	O (%)	
Rice husk	19.1	44.6	36.3	39.8	5.7	0.5	0.2	39.8	[2]
Cotton stalk	23.5	48.2	28.4	46.8	6.4	0.3	0.2	46.8	[2]
Rice straw	28.5	54.5	17.0	38.8	6.7	0.2	0.2	38.8	[2]
Wheat straw	25.5	49.3	25.2	41.7	5	0.4	0.3	41.7	[2]
Corn stalk	52.9	28.0	19.1	43.8	5.7	0.9	0.1	48.9	[3]
Corn cob	35.4	47.0	17.7	43.6	5.8	0.7	1.3	48.6	[3]
Elephant grass	34.3	31.5	34.2	44.5	5.4	1.4	—	31.8	[3]
Hazelnut shell	17.4	25.4	57.2	52.3	6.5	5.2	9.2	26.8	[3]
Sugarcane bagasse	33.1	42.7	24.2	45.1	6.05	0.3	—	42.7	[3]
Switch grass	51.6	32.3	16.1	44.7	5.7	0.3	—	49.1	[4]
Hazelnut husk	22.8	38.2	38.9	42.6	5.5	1.1	0.1	50.6	[5]
Walnut shell	25.4	23.5	51.1	47.5	6.3	0.4	—	45.6	[6]
Pinewood waste	28.4	43.0	28.5	49.3	6	0.04	—	44.5	[7]
Apple pomace	27.8	47.5	24.7	47.9	6.6	0.7	—	37.4	[8]
Chestnut shells	23.3	32.6	44.0	48.1	5.4	0.6	—	45.7	[9]
Cherry stones	28.0	28.1	43.9	51.1	7.2	3	—	38.6	[9]
Grape seeds	22.9	16.9	60.2	51.5	6.3	1.7	—	40.3	[9]
P. juliflora	18.8	51.6	29.6	43.3	6.32	1.3	0.07	48.9	[10]
Cashew nut shells	18.6	41.3	40.1	58.3	7	0.7	0.06	32	[3, 11]
Coconut shell	27.9	40.3	31.9	53.9	5.7	0.1	0.02	39.4	[3, 11]
Cagon grass	28.8	50.9	20.4	44.3	5.6	0.8	0.09	49.0	[12]
Karanja fruit Hull	48.4	12.0	39.6	45.1	6.1	—	0.36	48.4	[13]
Cotton stalk	23.5	48.2	28.4	46.8	6.4	0.3	0.2	46.8	[2]
Hibiscus rosa sinensis	19.1	44.6	36.3	40–43	4.7–6.0	0.8–4.9	0.04–0.7	34–38	[14, 15]
Nerium oleander	23.5	48.2	28.4	44	6.7	0.5–1.2	0.04–0.2	30–35	[14, 16]

**HEMI, hemicellulose; CEL, cellulose; LIG, lignin.*

Table 1.
Different biomasses, their constituents and elemental analysis.

17.4 to 52.9, the wt% of cellulose varies from 12 to 54.5% and the wt% of lignin varies from 16.1 to 60.2.

As shown in **Table 1**, there is significant variation across different biomasses. Thus, products resulting from biomass pyrolysis will also vary widely, and hence, it will be difficult to control the product profiles if enough attention is not paid to these chief constituents along with the biomass processing conditions.

This book chapter considers packed bed and fluidized bed pyrolyzers, which vary significantly in their heat transfer capabilities because of the mixing conditions. Section 2 of the chapter describes different pyrolysis processes and their associated product profiles. In view of their importance in pyrolysis, the reaction kinetics are provided in Section 3. These kinetic parameters are obtained from a combination of prior published work and the authors' experimental data, based on thermogravimetric analysis (TGA) and lumped parameter models. Section 4 deals with the modeling of continuous packed bed and fluidized bed pyrolyzers, accounting for the variation in product profiles due to variation in feed and operating parameters. Section 5 introduces the optimization of product profiles for these two types of pyrolyzers under continuous operation.

2. Different pyrolysis processes

The primary products of the pyrolysis are gas, oil, char and water. The ratios of these products depend on parameters such as particle size, heating rate, degradation temperature, and feed rate of the material [14, 17–22]. The different pyrolysis processes are flash/fast pyrolysis, intermediate pyrolysis, and slow pyrolysis [1, 17, 20, 23–25]. The pyrolysis process is classified based on the heating rate as well as the degradation temperature of biomass. The operating parameters of different pyrolysis processes are given in **Figure 1**.

2.1 Flash/fast pyrolysis

Flash/fast pyrolysis is performed with very high heating rates (~1000 K/s) for less than a second. Since, the entire biomass particles are required to be heated for a very short time, only particles of very small size (<1 mm) can be pyrolyzed using this method. Because of small particle size and very short residence times, the thermal gradients within the particles are small, and hence, there is low product variation. This method is preferable if the number of components in the product stream is required to be low. With most of the biomasses during fast pyrolysis, the products are usually non-condensable gases.

2.2 Medium/intermediate pyrolysis

Intermediate pyrolysis is performed at medium heating rates. This kind of pyrolysis is preferred for biomass in the particle size range of 1–5 mm. During the intermediate pyrolysis, there are huge thermal gradients within the particle itself. Hence, intermediate pyrolysis is preferred when different products are required from the same biomass. Further, this type of reactor does not require a separate unit for product separation, especially, oil and water.

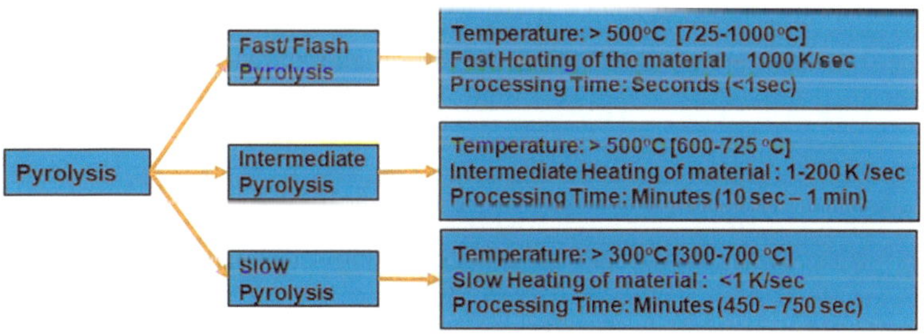

Figure 1.
Different types of pyrolysis and their operating conditions.

2.3 Slow pyrolysis

Slow pyrolysis is performed at very low heating rates ~0.1 K/s. The particle size during slow pyrolysis is between 2 and 10 mm. Slow pyrolysis is preferred if the charring characteristic of biomass is very high [4]. This is also the preferred type of pyrolysis when the residence time required during the pyrolysis is high.

Because of variations in the feed and the pyrolysis process, the thermal gradients are very high. Due to these thermal gradients within the particle and in the reactor, the reactions that the particles are undergoing are different, resulting in different product profiles. To design pyrolysis reactors with controlled product selectivity, it is important to understand which reactions are taking place in the reactor/biomass particle. This means that it is required to understand the kinetic parameters such as reaction order, activation energy, and pre-exponential factor for the reaction [14, 17, 23, 26–29]. Since, biomass particles are undergoing complicated reaction networks, the estimation of kinetic parameters were limited to the overall process in the early stages. Later, the kinetic parameters were estimated using the staged decomposition of biomass due to its chief constituents of biomass, namely, hemicellulose, cellulose, and lignin, and their relative proportions in a biomass [11, 14, 20, 30]. In the next section of this chapter, the kinetics of different biomasses are presented along with their corresponding parameters.

3. Pyrolysis kinetics and product profiles

Kinetics play an important role in the design of pyrolysis reactors. These kinetic parameters depend on the type of the biomass feed, mixing conditions in the reactor, and inter-particle & intra-particle thermal and mass transfer limitations. As mentioned earlier, the biomass pyrolysis consists of complicated reaction networks, making it difficult to analyze the pyrolysis kinetic parameters for individual reactions.

3.1 Pyrolysis kinetics

Different kinetic models were proposed by several authors working on the kinetics of the pyrolysis of biomass. Currently, there is no single model for the evaluation of kinetic parameters primarily based on the thermogravimetric (TGA) studies. The different types of kinetic models tested for the biomass pyrolysis are: (1) isoconversion (ISO) models such as Kissinger-Akahira-Sunose (KAS) model and Ozawa-Flynn-Wall (OFW) model, (2) generalized n^{th} order reaction model, (3) Coats-Redfern (CR) model, (4) three component mechanism with CR model, (5) Gaussian Distributed Activation Energy model (Gaussian DAEM), (6) Friedmann model, and (7) Starink model. The differences lie in their approach to relate them to rate of biomass degradation with kinetic parameters-mainly the activation energy and the pre-exponential factor.

The different models to analyze the pyrolysis thermal degradation data, were classified mainly into three categories namely, accelerating, decelerating, and sigmoidal model based on the shape of rate of degradation vs. time [17, 29]. The accelerating models are the ones whose rate of decomposition increases with the increase in time. For those biomasses which fall under this category a simple power law model with respect to rate of decomposition is more suitable [29]. The decelerating models are used whose rate of reaction decreases with increase in time, such as CR model and integral models (such as KAS and OFW models). This means, there are mass transfer limitations (diffusion limited) during the reaction due to the formation of products or inert layers that slows down the reaction. The sigmoidal models

Type of biomass	Model	Activation energy–E_a (kJ/mol)	Pre-exponential factor-A (min^{-1})	Reference
Rice husk	Gaussian DAEM	85–110	10^5–10^7	[31]
Cotton stalk	Direct method Integral method	Hemicellulose: 102.0 Cellulose: 98.5 Hemicellulose: 127.8 Cellulose: 72.5	1.22 0.45 2.6 0.35	[32]
Rice straw	Kissinger model OFW model KAS model	172.6 192.7 193.6	10^{11} 10^{22} 10^{15}	[32]
Wheat straw	nth-order reaction model	Cellulose and hemicellulose: 78 (n-0.65) Lignin: 80 (n-2.7)	10^7 10^6	[33]
Corn stalk	KAS model	62.7	10^7–10^8	[34]
Corn cob	CR model	64–80	10^3	[35]
Switch grass	KAS model	77.4	10^8–10^9	[34]
Elephant grass	Three component mechanism	Hemicellulose: 46.5–65.5 Cellulose: 108–127.2 Lignin: 45.6–53.5	10^4–10^6 10^8–10^{11} 10^2–10^3	[36]
Hazelnut husk	KAS model OFW model CR model	127.8 131.1 —	— — 10^5–10^6	[37]
Hazelnut shell	Friedman model KAS model DE Algorithm CR model	222.3 216.3 First zone: 35–153 Second zone: 20–135 35.9	— — 10^{-3}–10^8 10^{-4}–10^6 10^2	[38]
Walnut shell	Arrhenius model CR model	69.3 101.6	10^5 10^5	[6]
Pine wood	Arrhenius model	150	10^{11}	[39]
Apple pomace	Friedman model OFW model KAS model CR model	197.7 213.0 201.7 —	— — — 10^{-1}–10^{-3}	[8]
Chestnut shells	Friedman model KAS model OFW model Starink model DAEM	127.2–194.8 152.7–196.7 154.1–196.1 153.2–196.9 175.2	— — — — 10^{11}–10^{15}	[40]
Cherry stones	Arrhenius equation	Hemicellulose: 197.7 Cellulose: 213.0 Lignin: 201.7	10^{10} 10^7 10^2	[41]
Grape seeds	Gaussian DAEM Logistic DAEM	188 ± 3 190 ± 2	10^{-2} 10^{-2}	[42]
P. juliflora	KAS model OFW model Friedman model	204.0 203.2 219.3	10^{11} 10^5 10^{21}	[10]
Sugarcane bagasse	Direct method Integral method	Hemicellulose: 53.5, cellulose: 43 Hemicellulose: 87.7, cellulose: 77	10^{-1} 10^{-1} 10^{-1} 10^{-1}	[32]

Type of biomass	Model	Activation energy–E_a (kJ/mol)	Pre-exponential factor-A (min^{-1})	Reference
Cashew nut shells	CR model	Hemicellulose: 130.2 Cellulose: 174.3	10^{-1}–10^{-4} 10^{-2}–10^{-3}	[11]
Coconut shell	CR model	Hemicellulose: 179.6 Cellulose: 216.0	10^{-1}–10^{-5} 10^{-1}–10^{-3}	[11]
Imperata cylindrical (*Cagon grass*)	DAEM Global kinetic model	213.9 60–64	10^{13} 10^{1}–10^{2}	[12]
Karanja fruit Hull	KAS model OFW model	61.0 68.5	10^{6} 10^{6}	[13]
Hibiscus rosa sinensis	CR model	Hemicellulose: 55–91 Cellulose: 9–62 Lignin: 65–142	10^{6}–10^{8} 10^{-1}–10^{3} 10^{4}–10^{7}	[14]
Nerium oleander	CR model	Hemicellulose: 29–51 Cellulose: 11–43 Lignin: 71–109	10^{2}–10^{4} 10^{-1}–10^{3} 10^{2}–10^{5}	[14]

Table 2.
Kinetic parameters of pyrolysis of different biomasses.

represent the autocatalytic type reactions during biomass pyrolysis, such as OFW and KAS models. The choice of the kinetic model purely depends on the rate of decomposition vs. time curve.

The isoconversion model and Gaussian DAEM will give only the activation of energy for the overall thermal degradation process, which is further used in Arrhenius type model for finding the pre-exponential factor.

Table 2 provides the kinetic parameters reported by researchers for different biomasses that are widely used in pyrolysis.

Comparing **Tables 1** and **2**, for the biomasses with high hemicellulose content, the activation energy is between 50 and 65 kJ/kmol, and the pre-exponential factor is between 10^3 and 10^8 min^{-1}. The activation energy for high cellulose content is between 170 and 220 kJ/kmol and the pre-exponential factor is between 10^7 and 10^{21} min^{-1}. For the biomasses with high lignin content, the activation energy ranges from 65 to 140 kJ/kmol and the pre-exponential factor ranges from 10^{-1} to 10^7 min^{-1}. These wide variations in the kinetic parameters indicate that different reactions are occurring within the biomass particle.

3.2 Pyrolysis product profiles: lumped product distribution

It has been mentioned earlier that the products of biomass pyrolysis vary widely. There are not many studies aimed at intrinsic kinetic parameter estimation for each individual products of pyrolysis. The pyrolysis products are lumped based on their standard phase. The lumped reaction model for the biomass pyrolysis is given in Eqs. (1) and (2) [43, 44].

$$\text{Reaction 1}: \ Biomass \begin{array}{l} \rightarrow Gas \\ \rightarrow Tar/Liquids \\ \rightarrow Char \end{array} \tag{1}$$

The tar formed during the primary biomass decomposition shall further decompose into gas and char.

$$\text{Reaction 2}: \ Tar \begin{array}{l} \rightarrow Gas \\ \rightarrow Char \end{array} \tag{2}$$

Biomass type	Pyrolysis type	Pyrolysis conditions	Reactor type	Products (wt%)	Ref #
Rice husk	Intermediate	Temp.: >500°C HR: >3.33 K/s Size: <0.5 mm	Fluidized bed	Gas: 15% Liquids: 37–40% Char: 47%	[45]
	Slow	Temp.: >500°C HR: >0.33 K/s Size: <0.21 mm	Fixed bed	Gas: 32.7% Liquids.: 30.1% Char: 31.8%	[46]
Cotton stalk	Intermediate	Temp.: 490°C HR: 9 K/s Size: 1 mm	Fluidized bed	Gas: 23% Liquids: 36% Char: 16%	[47]
	Slow	Temp.: 600°C HR: 0.3 K/s Size: 3 mm	Fixed bed	Gas: 44.8% Liquids: 17.1% Char: 38%	[24, 48]
Rice straw	Intermediate	Temp.:500°C HR: 3.3 K/s Size: <0.5 mm	Fixed bed	Gas: 54% Liquids: 6% Char: 40%	[20]
	Slow	Temp.:600°C HR: 0.16 K/s Size: 0.3 mm	Fixed bed	Gas: 30% Liquids: 45% Char: 28%	[49, 50]
Wheat straw	Fast/flash	Temp.: 525°C HR: 250–1000 K/s Size: 0.25–1 mm	Fluidized bed	Gas: 26.9% Liquids: 34.9% Char: 28.0%	[28]
	Slow	Temp.: 450°C HR: 0.03 K/s Size: 1.33 mm	Fixed bed	Gas: 24% Liquids: 34% Char: 43%	[51]
Corn stalks	Intermediate	Temp.: 500°C HR: 8.3 K/s Size: <0.5 mm	Fluidized bed	Gas: 35% Liquids: 38% Char: 28%	[52]
	Slow	Temp.: 400°C HR: 0.83 K/s Size: <1 mm	Fixed bed	Gas: 12% Liquids: 50% Char: 38%	[53]
Corn cob	Fast/flash	Temp.: 550°C HR: 1000 K/s Size: 1–2 mm	Fluidized bed	Gas: 46% Liquids: 36% Char: 18%	[54]
	Intermediate	Temp.: 577°C HR: 10 K/s Size: 0.5–2 mm	Fixed bed	Gas: 49% Liquids: 28% Char: 23	[55]
Elephant grass	Fast/flash	Temp.: 480–520°C HR: 4000 K/s Size: 0.8–1.2 mm	Fluidized bed	Gas: 36.3% Liquids: 37.6% Char: 27.8%	[56, 57]
	Slow	Temp.: 500°C HR: 0.16 K/s Size: 0.21 mm	Fixed bed	Gas: 26% Liquids: 44.7% Char: 29.3%	[58]
Hazelnut shell	Slow	Temp.:550°C HR: 0.1 K/s Size: 0.85–1.8 mm	Fixed bed	Gas: 21.5 Liquids: 22.5% Water: 25% Char: 31%	[59]
	Intermediate	Temp.: 550°C HR: 5 K/s Size: 0.85–1.8 mm	Fixed bed	Gas: 27 Liquids: 34% Water: 11% Char: 28%	[59]
Sugarcane bagasse	Intermediate	Temp.: 500°C HR: 3.3 K/s Size: <0.5 mm	Fixed bed	Gas: 55% Liquids: 10% Char: 35%	[20]
	Slow	Temp.: 420°C HR: 0.35 K/s Size: <0.5 mm	Fixed bed	Gas: 24.4% Liquids: 43% Char: 32.6%	[60]

Biomass type	Pyrolysis type	Pyrolysis conditions	Reactor type	Products (wt%)	Ref #
Switch grass	Fast/flash	Temp.: 510°C HR: 1000 K/s Size: 0.25–1 mm	Fluidized bed	Gas: 16.5% Liquids: 57.9% Char: 20.0%	[28]
	Slow	Temp.: 500°C HR: 0.17 K/s Size: 0 < 1 mm	Fixed bed	Gas: 43.2% Liquids: 27.5% Char: 29.3%	[61]
Walnut shell	Intermediate	Temp.: 500°C HR: 5 K/s Size: 0.6–1.8 mm	Fixed bed	Gas: 15.6% Liquids: 31% Char: 27.5%	[62]
	Slow	Temp.: 550°C HR: 0.16–1 K/s Size: 0.075 mm	TGA	Gas: 10% Liquids: 25% Char: 40%	[63]
Pine wood	Fast/flash	Temp.: 400–500°C HR: >1000 K/s Size: 0.25–0.425 mm	Fluidized bed	Gas: 22% Liquids: 67% Char: 11%	[64]
	Intermediate	Temp.: 500°C HR: 5 K/s Size: 06–0.85 mm	Fixed bed	Gas: 26% Liquids: 43% Char: 23% Water: 14%	[65]
	Slow	Temp.: 700°C HR: 0.16 K/s Size: <1 mm	Fixed bed	Gas: 25% Liquids: 58% Char: 18%	[66]
Apple pomace	Slow	Temp.: 400°C HR: 0.08–0.3 K/s Size: 420–840 μm	Fixed bed	Gas: 71.5% Liquids: 25.4% Char: 3%	[23]
Coconut shell	Intermediate	Temp.: 500°C HR: 3.3 K/s Size: <0.5 mm	Fixed bed	Gas: 51% Liquids: 10% Char: 39%	[20]
	Slow	Temp.: 550°C HR: 1 K/s Size: 1.18–1.8 mm	Fixed bed	Gas: 30–33% Liquids: 38–44% Char: 22–31%	[67]
Cashew nut shells	Slow	Temp.: 400–450°C HR: 0.166 K/s Size: 0.25 mm	Fixed bed	Gas: 57.4% Liquids: 23.5% Char: 19.1%	[68]
Cherry stones	Slow	Temp.: 400–500°C HR: 0.083–0.33 K/s Size: 0.32–2 mm	TGA	Gas: 8.8–47.6% Liquids: 32–58% Char: 20–56.8%	[41]
P. juliflora	Fast/flash	Temp.: 450°C HR: >4000 K/s Size: 0.25–0.5 mm	Fluidized bed	Gas: 12.5% Liquid: 62.5% Char: 25%	[57, 69]
	Slow	Temp.: 600°C HR: 0.33 K/s Size: 0.2–0.5 mm	Fixed bed	Gas: 44 Liquid: 38.3% Char: 36.8%	[70]
Cogon grass	Slow	Temp.: 500°C HR: 0.36 K/s Size: 0.25–1 mm	Fixed bed	Gas: 49.1–74.1% Oil: 3.2–20.8% Char: 22.6–30.5%	[12]

Temp., temperature; HR, heating rate; size, biomass particle size.

Table 3.
Product profiles from pyrolysis of different biomass in fixed and fluidized bed pyrolyzers.

For practical purposes, the products are classified into gases, liquids, and char. The lumped product distribution from pyrolysis of different biomasses, different pyrolysis methods, pyrolysis conditions and two different pyrolyzers (fixed bed and fluidized) are provided (**Table 3**).

From the data presented in **Table 3**, it can be generalized that the liquid products will be more in fluidized bed reactors, whereas more char is seen in fixed bed reactors. Further, slow pyrolysis results in more gas and char when compared to fast pyrolysis. This might be due to the liquid products formed in reaction 1, shown in Eq. (1), being further consumed by reaction 2, shown in Eq. (2) resulting in more gas and char. The relative proportions of these products depend on various parameters mentioned earlier.

4. Process modeling of pyrolysis reactors

The biomass pyrolysis has been mainly carried out in two types of reactors: fixed bed (or packed bed) reactors and fluidized bed reactors, by many authors, as presented in **Table 3**. The process models for these two reactors are developed from fundamental laws of conservation of mass and energy along with empirical relations for properties such as specific heat, density, diffusivity, mass and heat transfer coefficients, etc. The packed bed pyrolyzers are further classified as down-draft and updraft pyrolysis reactors. The detailed comparison of the general packed bed and fluidized bed reactors is given in **Table 4**.

The biomass usually contains moisture. The moisture needs to be removed before the pyrolysis stage. If this is carried to the pyrolysis stage, the gasification reactions such as steam reformation may kick off resulting in undesirable products.

Gasifier type	Specifications/conditions
Updraft fixed pyrolyzer	• The biomass is fed from the top of the pyrolyzer, and the inert gas (if any) fed from bottom. • Char resulting from pyrolysis falls down and may accelerate the pyrolysis. • The pyrolysis gases and along with the liquid tar (in the form of gas) leaves from the top of the pyrolyzer. • The ash (inert component of the biomass) is collected at the bottom of the gasifier. • Operating temperature ranges from 300 to 750°C.
Downdraft fixed bed pyrolyzer	• The biomass is fed from top of the pyrolyzer along with inert gas allowing the feed and gases move in the same direction. • The feed is broken down, falling down the gasifier under gravity. A bed of hot char through which the gases are allowed to pass through (a secondary reaction zone) ensures the pyrolysis products travelling from top are further broken down. This increases the residence time through the pyrolysis stage. An exit for the pyrolysis products is provided just above the bottom of the pyrolyzer. • The ash collected under the grate at the bottom the pyrolyzer. • Operating temperature ranges from 300 to 750°C.
Fluidized bed reactor	• A bed of fine inert solid material is present at the bottom of the pyrolyzer. The inert gas is fed from the bottom of the pyrolyzer fast enough (1–3 m/s) to agitate the material. • The biomass feed is fed in from the side, mixes with the inert gas and the products of the pyrolysis leave from the top. • The operating temperature is below 900°C to avoid ash melting and sticking to the wall.

Table 4.
Comparison of fixed bed and fluidized bed pyrolyzers.

The modeling of the packed bed and the fluidized bed pyrolyzers is discussed in detail below.

The major difference between the updraft pyrolyzer and the downdraft pyrolyzer is only the inlet and outlet locations, which changes the residence time and mixing conditions. For the purpose of the modeling of biomass pyrolysis and optimization of product profiles, the updraft and downdraft pyrolyzers are treated as fixed bed pyrolyzers.

4.1 Modeling of packed bed pyrolysis reactor

The modeling of the packed bed pyrolysis reactors is divided into drying stage and the pyrolysis stage. The modeling of these two stages has been is presented in detail below.

4.1.1 Modeling of drying stage in packed bed pyrolyzers

The drying of biomass particles will happen in two stages: (a) constant rate period and (b) falling rate period. The rate of drying in these two stages is given separately by the following equations.

The rate of drying during the constant rate period is given by [71]

$$\frac{dX}{dt} = -k_c = 1.3 \times 10^{-9} T_g^{4.112} v_g^{0.219} \tag{3}$$

The rate of drying during the falling rate period is given by [71]

$$\begin{aligned} \frac{dX}{dt} &= -K\left(X - X_{eq}\right) \\ K &= 0.011 \exp\left(-201.8/T_g\right) \end{aligned} \tag{4}$$

Eqs. (3) and (4), provide an estimation of rate of drying in both constant and falling rate periods, with rate of drying depending on the particle temperature. During the pyrolysis process, the biomass particle temperature depends on the rate of heat transfer between the gas and the particle. This rate of heat transfer depends on the flow characteristics involving Reynolds and Prandtl number. For 20 < Re < 1000, the heat transfer coefficient is given by [71, 72]

$$\begin{aligned} h &= 3.26 C_{pg} G_g \mathrm{Re}^{-0.65} \mathrm{Pr}^{2/3} \\ \mathrm{Re}_p &= \frac{\rho_{gas} U_o d_p}{\mu} \end{aligned} \tag{5}$$

where k_c is the mass transfer coefficient at constant drying period in s^{-1}; K is the mass transfer coefficient at falling drying period in s^{-1}; Re is the Reynolds number of gas; Pr is the Prandtl number of gas; T_g is the gas temperature in °C; v_g is the gas velocity in m/s; C_{pg} is the heat capacity of gas in J/(kg K); ε_{mf} is the porosity at minimum fluidization; X is the moisture content in the biomass; X_{eq} is the moisture content at the end of constant rate period; G_g is the gas mass flux in kg/(m^2 s).

4.1.2 Modeling of pyrolysis in packed bed reactors

Eqs. (1) and (2) represents the reactions reported during pyrolysis of biomass with lumped product approach. Since the reaction presented in Eq. (1) involves only the solid biomass as reactant, most of the reactions are treated as first order reactions using Arrhenius type model [73]. The Arrhenius model parameters for the lumped reaction model that produces gaseous, liquid, and char for different constituent of biomass are given in **Table 5**.

The volume and the moles of the products of the pyrolysis zone was found with the help of the plug flow reactor (PFR) design equation.

$$\frac{V}{F_{Biomass}} = \int_0^{X_f} \frac{dX_{Biomass}}{-r_{Biomass}} \tag{6}$$

$$\left(-r_{Biomass} = r_{Gas} + r_{Liquids} + r_{Char}\right)$$

where V is the volume of the pyrolyzer in l, $F_{Biomass}$ is the molar feed rate of biomass in mol/s; X_f is the desired biomass conversion; $-r_{Biomass}$ is the rate of biomass consumption in mol/(lit. s); r_{Gas} is the rate of formation of gaseous products in mol/(lit. s); $r_{Liquids}$ is the rate of formation of liquid products in mol/(lit. s); r_{Char} is the rate of formation of char in mol/(lit. s).

Since, the pyrolysis stage is non-isothermal in nature, it is essential to model the temperature along the length of the pyrolyzer which involves external heating. The energy balance equation was obtained based on the shell energy balance to find the temperature profile along the length of the pyrolysis zone. The steady state energy balance equation results in the following differential equation:

$$k\frac{d^2T}{dr^2} - \rho C_p v_z \frac{dT}{dz} + \frac{\varepsilon_R}{R}\sigma\left(T_W{}^4 - T^4\right) - \Delta H_R F_{AO} dX_A = 0 \tag{7}$$

where k is the thermal conductivity in W/(m. K); ρ is the density in kg/m^3; C_p is the specific heat capacity in J/(kg. K); v_z is the gas velocity in m/s; ε_R is the emissivity of the body during radiation; σ is the Stephan-Boltzmann constant = 5.67E−08 W/(m^2 K^4); R is the radius of oxidation zone in m; T_W is the source temperature in K; T is the reactor temperature in K; r, z are the radial and axial directions.

Since, Eq. (7) is combined convection diffusion equation with heat source (heat of reaction) and external heat supply, two boundary conditions are required: one in radial direction and one in axial direction.

The boundary conditions for solving the energy balance equation are:

$$@r = R, k\frac{dT}{dr} = Radiation\ flux$$

$$@r = 0; k\frac{dT}{dr} = 0\ (Tempearature\ minimum\ condition)$$

$$@z = 0; T = T_o$$

Pyrolysis product →	Gas		Tar/liquids		Char	
Biomass component ↓	**A (s^{-1})**	**E$_a$ (J/mol)**	**A (s^{-1})**	**E$_a$ (J/mol)**	**A (s^{-1})**	**E$_a$ (J/mol)**
HEMI*	2.1×10^{16}	1.8×10^5	8.7×10^{14}	2.0×10^5	2.6×10^{11}	1.5×10^5
CEL*	2.8×10^{19}	2.4×10^5	3.3×10^{14}	1.9×10^5	1.3×10^{10}	1.5×10^5
LIG*	9.6×10^8	1.1×10^5	1.5×10^9	1.4×10^5	7.7×10^6	1.1×10^5

**HEMI, hemicellulose; CEL, cellulose; LIG, lignin.*

Table 5.
Kinetic parameters for lumped models for biomass pyrolysis.

4.2 Modeling of fluidized bed pyrolysis reactor

As mentioned in the modeling of the packed bed pyrolysis, the pyrolysis process occurs after the drying of the biomass.

4.2.1 Modeling of drying stage in fluidized bed pyrolyzers

The equations for the drying stage in a fluidized bed pyrolyzers are similar to those in a packed bed pyrolyzer, as provided in Section 4.1.1. Because there is vigorous mixing in the fluidized bed pyrolysis, the extent of heat and mass transfer is very high. Kunii and Levenspiel confirmed that there is more than one phase during bubbling fluidization [74]. These two phases are named as bubble phase (primarily the fluid used as fluidizing the medium) and emulsion phase (mixture of biomass particles and fluid). The interface between these two phases is named as "cloud." The concept of two-phase model along with the mass transfer between the two phases is depicted in **Figure 2**. It is to be noted that the primary reactions during biomass pyrolysis shall occur only in the solid phase (i.e., in the emulsion phase) and the secondary reactions occur in both phases.

The most widely used Kunii and Levenspiel model (K-L model) expresses the overall heat and mass transfer coefficient in a bubbling fluidized bed considering the resistance for heat and mass transfer between the bubble-cloud interface and resistance for heat and mass transfer between emulsion and cloud [74]. The cloud-bubble interface heat and mass transfer are functions of gas velocity and conduction/diffusion from a thin cloud layer into the bubble. The emulsion and cloud transfer are only due to conduction/diffusion between the emulsion phase and the cloud boundary. They had also suggested additional mass transfer resulting from particles dispersed in the bubbles. However, recent advanced imaging technique, have shown bubble free particles in most cases.

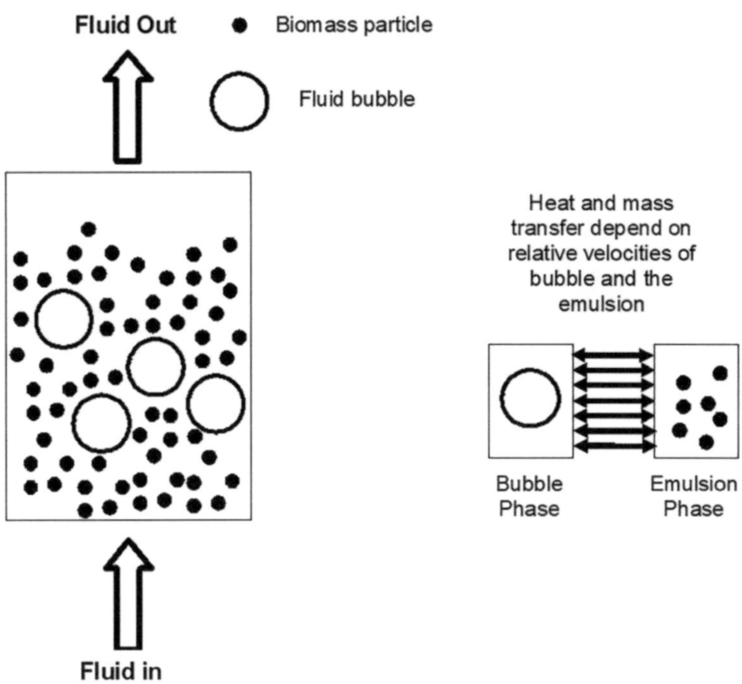

Figure 2.
The two-phase model for fluidized bed pyrolyzer.

Based on the two-phase model, the overall mass transfer coefficient is estimated using Eqs. (8)–(10) [75].

Cloud-bubble mass transfer coefficient

$$k_{cb} = 1.5\frac{U_{mbf}}{d_b} + 5.85\left(\frac{D_{AB}{}^{0.5}g^{0.25}}{d_b^{1.25}}\right) \tag{8}$$

Dense-cloud mass transfer coefficient

$$k_{dc} = 6.77\left(\frac{D_{AB}\varepsilon_{mf}U_b}{d_b^3}\right) \tag{9}$$

Overall mass transfer coefficient (in s^{-1})

$$\frac{1}{k_{overall}} = \frac{1}{k_{cb}} + \frac{1}{k_{dc}} \tag{10}$$

The heat transfer coefficient is estimated using the correlation given in Eq. (11) [76].

$$h_c = 0.15\frac{k_g}{d_p}\mathrm{Re}_p{}^{0.35}Ar^{0.25}\left(V_{gas} \le U_o\right) \tag{11}$$

$$\mathrm{Re}_p = \frac{\rho_{gas}U_o d_p}{\mu} \tag{12}$$

$$Ar = \frac{d_p^3\rho_{gas}\left(\rho_{solid} - \rho_{gas}\right)g}{\mu^2} \tag{13}$$

where k_{dc} is the mass transfer coefficient between dense emulsion and cloud in m/s; k_{cb} is the mass transfer coefficient between bubble and cloud in m/s; $k_{overall}$ is the overall mass transfer coefficient between bubble and cloud in m/s; h_c is the overall heat transfer coefficient in $W/(m^2\ K)$; Re_p is the Particle Reynolds number; Ar is the Archimedes number; U_{mbf} is the bubble velocity at minimum fluidization in m/s; D_{AB} is the binary diffusivity in m^2/s; U_b is the bubble velocity in m/s; ε_{mf} is the bed voidage at minimum fluidization; ρ_{solid} is the density of biomass particle in kg/m^3; ρ_{gas} is the density of fluidizing gas in kg/m^3; d_b is the bubble diameter in m.

4.2.2 Modeling of pyrolysis stage in fluidized bed pyrolyzers

Due to vigorous mixing in the fluidized beds, the following assumptions are made for modeling of the pyrolysis stage in the fluidized bed reactors:

1. Isothermal operation of the reactor,

2. No radial variations, and

3. Reactions occurring only in the solid phase (only in the emulsion phase).

Based on these assumptions, the following models (separately for emulsion phase and bubble phase) can be obtained from the fundamental component balance:

$$\text{Bubble phase}: \; \varepsilon u_b \frac{dC_{A,b}}{dz} - k_L(C_{A,b} - C_{A,e}) = 0 \tag{14}$$

$$\text{Emulsion phase}: \; (1-\varepsilon)u_e \frac{dC_{A,e}}{dz} + k_L(C_{A,b} - C_{A,e}) - (1-\varepsilon)D_{eff} \frac{d^2C_{A,e}}{dz^2} + (-r_A)\rho_e(1-\varepsilon) = 0; \tag{15}$$

where b is the bubble phase and e is the emulsion phase.

5. Optimization of product profiles in biomass pyrolysis

As discussed in Section 1, since the product profiles of biomass pyrolysis are known to greatly depend on various parameters, it is necessary to determine the parameters that have the largest effect on the ratios of important lumped products discussed in Section 2.

5.1 Response surface optimization methodology: the mixture design

The following quantifiable factors were chosen for the optimization.
The biomass-based factors are:

a. Hemicellulose fraction

b. Cellulose fraction

c. Lignin fraction

The process-based parameters have been divided into two categorical type parameters.

a. The type of pyrolysis-slow and fast/flash pyrolysis

b. The type of reactor-fixed bed and fluidized bed reactors

The variation of constituent composition of biomass is obtained from **Table 1**. The amounts of gaseous, liquid, and products for each biomass for different process parameters and different reactor types are obtained from **Table 3**.

Since hemicellulose, cellulose, and lignin are biomass constituents, the sum of their fractions must be equal to one. Accordingly, a mixture design was chosen for the statistical design of the experiments.

A total of 52 data sets were selected combining **Tables 1** and **3** (based on the variations in mixture design), and the experiments conducted and data analyzed.

The depletion of fossil fuels has created interest in obtaining the fuels from alternate sources such as biomass, especially for transportation fuels. Therefore, the objective of the product profiles from biomass pyrolysis is aimed to maximize the liquid (tar) products and simultaneously minimize the production of gas and char. The reason for the minimization of gaseous products is that they may contain carbon monoxide and carbon dioxide which are of less calorific value when compared to the hydrocarbons.

Ternary diagrams are common in representing factor levels in a mixture design. In such a diagram, with three factors (x1, x2, and x3), the vertex represents a pure component (i.e., x1 = 100%) while its opposite edge holds a value of x1 = 0 (with x2 + x3 = 100%). The same holds for the other two vertices and edges.

However, it may be noted that in certain experiments the sum of proportions of the three components may be deliberately constrained to equal a specific value. This feature has been used in our experimental design. For example, referring to **Figure 3**, the three vertices represent hemicellulose, cellulose and lignin. The maximum fraction of these three constituents is 0.764, 0.742 and 0.7444 for hemicellulose, cellulose and lignin, respectively. The 0.119, 0.117, and 0.139 marked on the sides of the triangle represent the minimum values for the biomass constituents.

5.2 Optimization of gas yield

The analysis of minimization of gas yield is separated into two types: (a) fast pyrolysis in fluidized bed pyrolyzer and (b) slow pyrolysis in fixed bed pyrolyzer. These two were chosen to account for the extremities of the process conditions. The contour plot for fast pyrolysis in fluidized bed is shown in **Figure 3**. The three vertices of the triangle are the maximum points of hemicellulose, cellulose, and lignin. For the minimization of gaseous products during fast pyrolysis in fluidized bed, the hemicellulose content in biomass is required to be high, when compared to the cellulose and lignin. More than 40% hemicellulose and ~20% of each of lignin and cellulose in the biomass feed will minimize the gas yield during fast pyrolysis in fluidized bed. This is easily achievable, because there are biomasses with hemicellulose content higher than 40% (please refer to **Table 1**). Considering fast pyrolysis of wheat straw in fluidized bed reactor, which contains 25.5% hemicellulose, 49.3% cellulose, and 25.2% of lignin, the gas yield is ~30%, which is consistent with the experimentally obtained data presented in **Table 3**.

The gas yield during slow pyrolysis in fixed bed pyrolysis is higher than that of fast pyrolysis in fluidized bed reactors. This is due to the fact that the residence time in the fixed bed reactors is higher when compared to that of fluidized bed reactors. Due to longer residence times, there are secondary reactions, that is, the conversion of liquid into gas and char. The optimal minimum gas yield during slow biomass pyrolysis in fixed bed reactors is ~16%, as shown in contour plot in **Figure 4**. The minimum gas yield can be obtained at various conditions. Specifically, it is obtained either at high hemicellulose content or at medium cellulose and lignin content in the

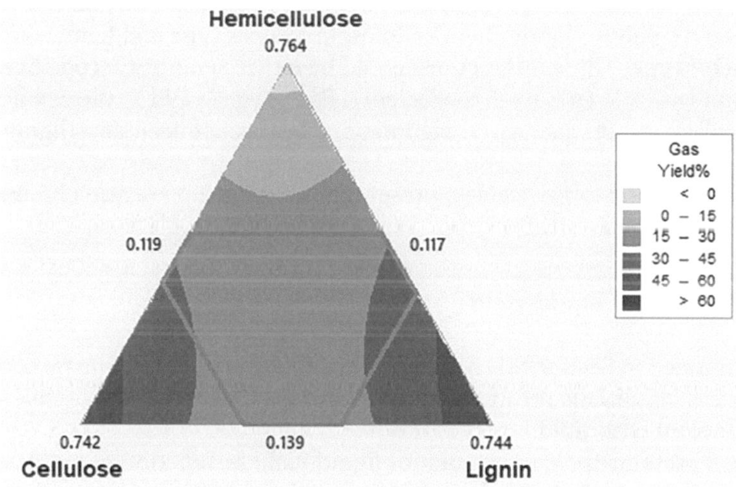

Figure 3.
Contour plot of gas yield in biomass pyrolysis for fast pyrolysis in fluidized bed reactor.

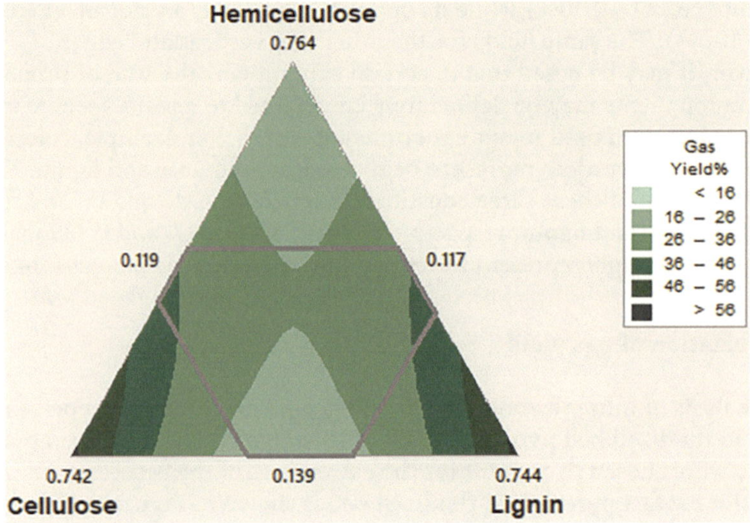

Figure 4.
Contour plot of gas yield in biomass pyrolysis for slow pyrolysis in fixed bed reactor.

biomass. Considering the wheat straw decomposition during slow pyrolysis in fixed bed, the gas yield is 16–26% as given by **Figure 4**.

The gas yield during biomass pyrolysis may be obtained from statistical modeling, presented in Eq. (16). This equation was obtained after removing the model terms that were not statistically significant, that is, terms with p-value < 0.05.

$$GasYield(\%) = -20.4^*H + 108.8^*C + 111.3^*L - 377.8^*C^*L - 18.3^*H^*P + 50.1^*H^*L^*R \quad (16)$$

where H is the hemicellulose fraction in biomass; C is the cellulose fraction in biomass; L is the lignin fraction in biomass; P is the pyrolysis type (fast or slow); R is the reactor type (fluidized bed or fixed bed).

The coefficients present in Eq. (16) represents whether the effect is positive or negative. The coefficient for hemicellulose is negative. This means, the higher the hemicellulose content, lesser will be the gas yield. From Eq. (16), there are interactions between cellulose-lignin, hemicellulose-pyrolysis type and hemicellulose-lignin-reactor type. These interactions could be either synergistic (positive coefficient) or antagonistic (negative coefficient). For the gas yield%, the cellulose-lignin and hemicellulose-pyrolysis are antagonistic, whereas hemicellulose-lignin-reactor type is synergistic.

Equations similar to gas yield%, presented in Eq. (16) were also obtained for liquid yield% and char yield%.

5.3 Optimization of liquid yield

As mentioned in Section 5.1, it is important find the level of biomass constituents and operating conditions for pyrolysis to maximize the liquid yield, while minimizing both gas and char yield.

Figure 5 presents the contour plot of liquid yield as function of biomass constituents during fast pyrolysis. The maximum liquid yield may be obtained at high hemicellulose, low lignin and low cellulose content. For biomasses whose cellulose content is high, the liquid yields are very low and for the biomasses whose lignin

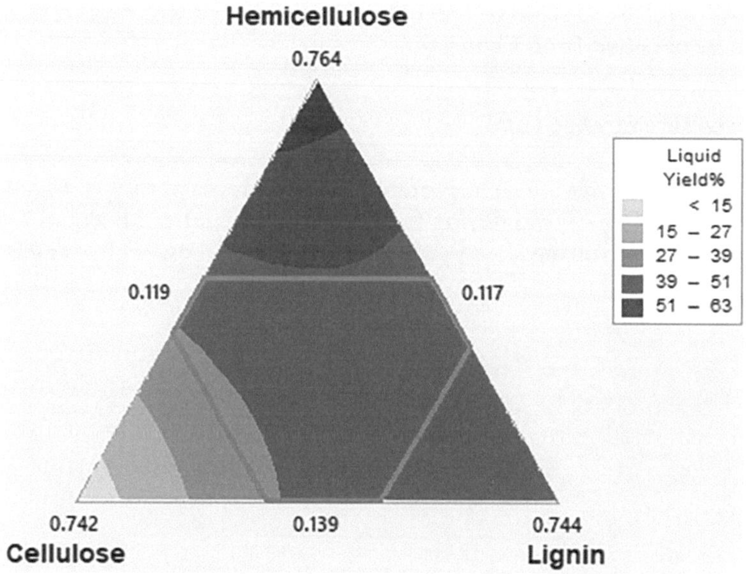

Figure 5.
Contour plot of liquid yield in biomass pyrolysis for fast pyrolysis in fluidized bed reactor.

content is high, the liquid yield ranges from 40 to 60%. For example, one of the biomasses containing high hemi cellulose is switch grass, for which, if decomposed during fast pyrolysis in fluidized bed, the liquid yield is more than 51% (from **Figure 5**), which is consistent with the experimentally obtained data.

Figure 6 shows the liquid yield during slow pyrolysis in fixed bed pyrolyzes. Comparing **Figure 5** with **Figure 6**, it can be concluded that, the liquid yield is higher in fast pyrolysis in fluidized bed than slow pyrolysis in fixed bed. Further, the maximum liquid yield that can be obtained is ~40%. This can be obtained only at high hemicellulose content, medium cellulose content and low lignin content. The liquid yield for the high hemicellulose content biomasses, such as switch grass,

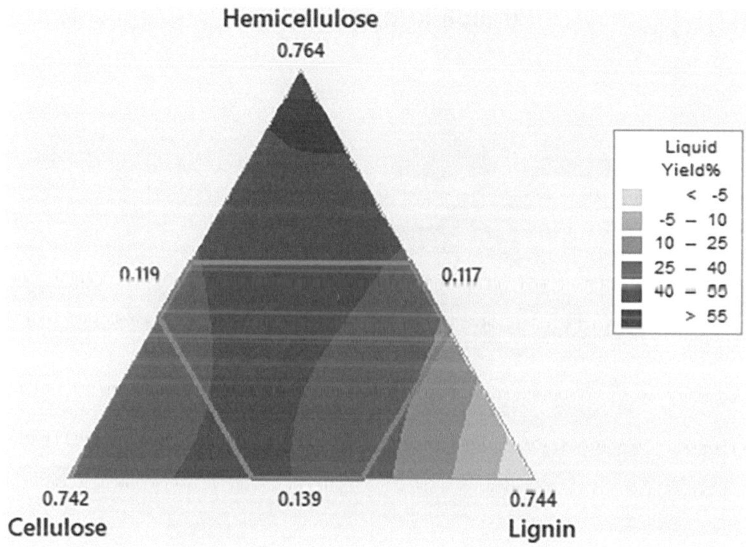

Figure 6.
Contour plot of liquid yield in biomass pyrolysis for slow in fixed bed reactor.

during slow pyrolysis in fixed bed reactor, the liquid yield will vary from 25 to 40%, which can be obtained from **Figure 6**.

5.4 Optimization of char yield

Minimization of char is also important during biomass pyrolysis. **Figure 7** shows the contour plot of char yield during fast pyrolysis in fluidized bed and **Figure 8** shows the char yield during slow pyrolysis in fixed bed. Lower char yields may be

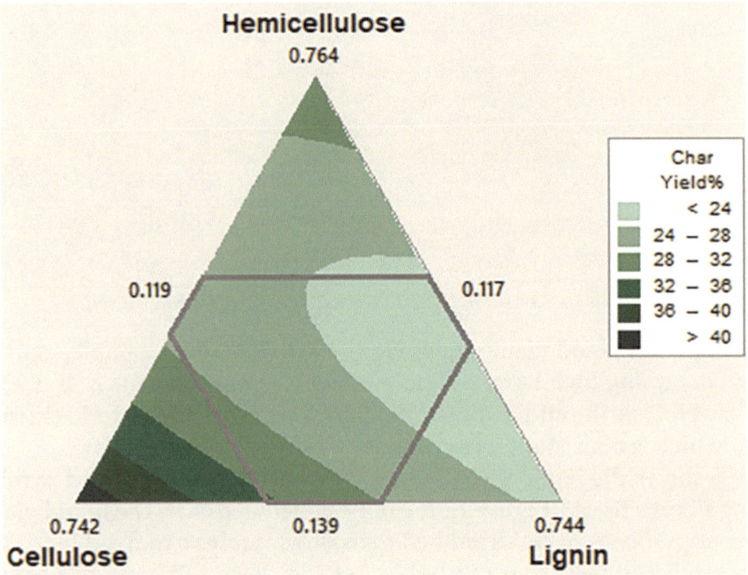

Figure 7.
Contour plot of char yield in biomass pyrolysis for fast pyrolysis in fluidized bed reactor.

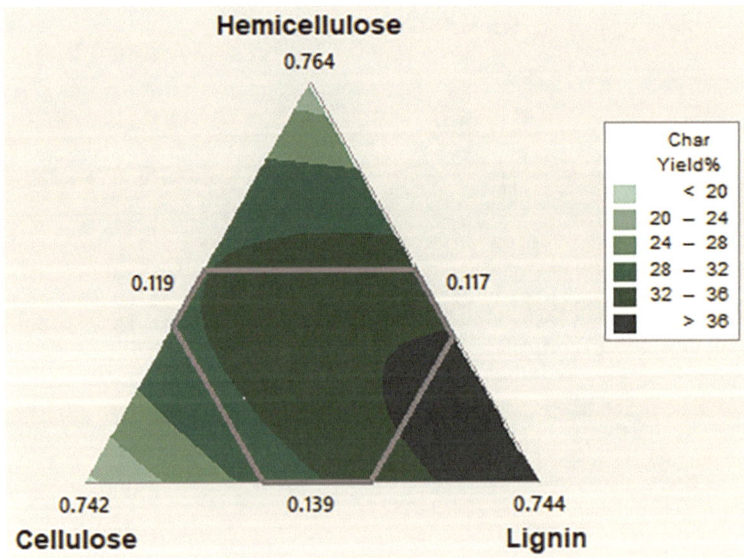

Figure 8.
Contour plot of char yield in biomass pyrolysis for slow pyrolysis in fixed bed reactor.

obtained during fast pyrolysis in fluidized bed from biomasses containing higher amount of lignin, medium amounts of hemicellulose and lower amounts of cellulose (lignocellulosic biomass in other words). For example, the high lignin biomasses such as hazelnut shell will yield less char during fast pyrolysis in fluidized bed reactor.

Low char yields may be obtained for biomasses with high hemicellulose and high lignin during slow pyrolysis in fixed bed operation.

6. Conclusions

Biomasses vary widely in their chief constituents namely, hemicellulose, cellulose, and lignin. The order of the degradation temperature for the three chief constituents is hemicellulose < cellulose < lignin. For obtaining controlled and useful product profiles such as liquid fuels via pyrolysis, it is important to understand the pyrolysis process conditions, such as heating rate and temperature. For optimization, the pyrolysis operating parameters are classified as slow and fast pyrolysis and the reactors are classified as fluidized bed and fixed bed. The data required for optimization was obtained from many sources, including experimental data for various biomasses and process conditions. The response surface optimization was performed using mixture design with three levels for biomass constituents and two levels each for pyrolysis type and reactor type. The objective of this book chapter is to model the reaction using lumped (gas, liquids, and char) approach and find the feed and operating parameters to maximize the liquid yield. In general, the liquid yield was found to be higher during fast pyrolysis/fluidized bed reactors when compared to slow pyrolysis/fixed bed reactors. This is due to the large residence time in the slow pyrolysis/fixed bed reactors. The maximum yield may be obtained for the biomasses containing high hemicellulose such as switch grass and corn stock.

The authors recommend further optimization with respect to particle size, heating rate, and biomass feed rate.

Acknowledgements

The authors acknowledge the financial support from the Center of Excellence in Advanced Materials & Green Technologies, granted under the Frontier Areas of Science and Technology (FAST) scheme to establish the Centers of Excellence by Ministry of Human Resource Development (MHRD), Government of India.

Conflict of interest

There is no conflict of interest.

Notes/Thanks/Other declarations

The authors would like to thank the students who attended the course "Waste to Energy part of Masters in Materials Science and Engineering program, Amrita Vishwa Vidyapeetham," who had helped in obtaining the data and supported the work through discussions.

Author details

Udaya Bhaskar Reddy Ragula*, Sriram Devanathan and Sindhu Subramanian
Department of Chemical Engineering and Materials Science, Amrita Vishwa Vidyapeetham, Coimbatore, India

*Address all correspondence to: u_bhaskarreddy@cb.amrita.edu

References

[1] Basu P. Biomass Gasification and Pyrolysis Handbook. Kidlington, Oxford, UK: Elsevier; 2010. DOI: 10.1016/B978-0-12-374988-8.00001-5

[2] Raj T, Kapoor M, Gaur R, Christopher J, Lamba B, Tuli DK, et al. Physical and chemical characterization of various indian agriculture residues for biofuels production. Energy & Fuels. 2015;**29**:3111-3118. DOI: 10.1021/ef5027373

[3] Dhyani V, Bhaskar T. A comprehensive review on the pyrolysis of lignocellulosic biomass. Renewable Energy. 2018;**129**:695-716. DOI: 10.1016/j.renene.2017.04.035

[4] Jahirul MI, Rasul MG, Chowdhury AA, Ashwath N. Biofuels production through biomass pyrolysis—A technological review. Energies. 2012;**5**: 4952-5001. DOI: 10.3390/en5124952

[5] Braga RM, Melo DMA, Aquino FM, Freitas JCO, Melo MAF, Barros JMF, et al. Characterization and comparative study of pyrolysis kinetics of the rice husk and the elephant grass. Journal of Thermal Analysis and Calorimetry. 2014;**115**:1915-1920. DOI: 10.1007/s10973-013-3503-7

[6] Uzun BB, Yaman E. Pyrolysis kinetics of walnut shell and waste polyolefins using thermogravimetric analysis. Journal of the Energy Institute. 2016;**90**:1-13. DOI: 10.1016/j.joei.2016.09.001

[7] Räsänen T, Athanassiadis D. Basic chemical composition of the biomass components of pine, spruce and birch. Forest Refine. 2013;**3**:1-4

[8] Guerrero MRB, Marques Da Silva Paula M, Zaragoza MM, Gutiérrez JS, Velderrain VG, Ortiz AL, et al. Thermogravimetric study on the pyrolysis kinetics of apple pomace as waste biomass. International Journal of Hydrogen Energy. 2014;**39**:16619-16627. DOI: 10.1016/j.ijhydene.2014.06.012

[9] Özsin G, Pütün AE. Kinetics and evolved gas analysis for pyrolysis of food processing wastes using TGA/MS/FT-IR. Waste Management. 2017; **64**:315-326. DOI: 10.1016/j.wasman.2017.03.020

[10] Chandrasekaran A, Ramachandran S, Subbiah S. Determination of kinetic parameters in the pyrolysis operation and thermal behavior of *Prosopis juliflora* using thermogravimetric analysis. Bioresource Technology. 2017; **233**:413-422. DOI: 10.1016/j.biortech.2017.02.119

[11] Tsamba AJ, Yang W, Blasiak W. Pyrolysis characteristics and global kinetics of coconut and cashew nut shells. Fuel Processing Technology. 2006;**87**:523-530. DOI: 10.1016/j.fuproc.2005.12.002

[12] Li LL, Fu XB, Wang XN, Tian YY, Qin S. Pyrolytic characteristics and kinetic studies of agricultural wastes-four kinds of grasses. Energy Sources, Part A: Recovery, Utilization, and Environmental Effects. 2016;**38**: 1156-1162. DOI: 10.1080/15567036.2013.856967

[13] Islam MA, Asif M, Hameed BH. Pyrolysis kinetics of raw and hydrothermally carbonized Karanj (*Pongamia pinnata*) fruit hulls via thermogravimetric analysis. Bioresource Technology. 2015;**179**:227-233. DOI: 10.1016/j.biortech.2014.11.115

[14] Subramanian S, Ragula UBR. Pyrolysis kinetics of *Hibiscus rosa sinensis* and *Nerium oleander*. Biofuels. 2018; **7269**:1-15. DOI: 10.1080/17597269.2018.1432274

[15] Vinodhini S, Malathy NS. Bioprospecting of plants fibre of Coimbatore district of Tamil Nadu. International Journal of Plant Sciences. 2009;**4**:444-445

[16] Jabli M, Tka N, Ramzi K, Saleh TA. Physicochemical characteristics and dyeing properties of lignin-cellulosic fibers derived from *Nerium oleander*. Journal of Molecular Liquids. 2018;**249**: 1138-1144. DOI: 10.1016/j.molliq. 2017.11.126

[17] Suriapparao DV, Vinu R. Effects of biomass particle size on slow pyrolysis kinetics and fast pyrolysis product distribution. Waste and Biomass Valorization. 2017;**9**:1-13. DOI: 10.1007/ s12649-016-9815-7

[18] Gaston KR, Jarvis MW, Pepiot P, Smith KM, Frederick WJ, Nimlos MR. Biomass pyrolysis and gasification of varying particle sizes in a fluidized-bed reactor. Energy & Fuels. 2011;**25**: 3747-3757. DOI: 10.1021/ef200257k

[19] Niu Y, Tan H, Liu Y, Wang X, Xu T. The effect of particle size and heating rate on pyrolysis of waste capsicum stalks biomass. Energy Sources, Part A: Recovery, Utilization, and Environmental Effects. 2013;**35**: 1663-1669. DOI: 10.1080/ 15567036.2010.509084

[20] Tsai WT, Lee MK, Chang YM. Fast pyrolysis of rice straw, sugarcane bagasse and coconut shell in an induction-heating reactor. Journal of Analytical and Applied Pyrolysis. 2006; **76**:230-237. DOI: 10.1016/j.jaap. 2005.11.007

[21] Lignite M. Effects of heating rate and particle size on pyrolysis kinetics of mengen lignite. Energy Sources. 2001;**23**:337-344. DOI: 10.1080/ 009083101300110887

[22] Basu P. Biomass Gasification and Pyrolysis: Practical Design and Theory. Kidlington, Oxford, UK: Academic Press; 2010. https://doi.org/10.1016/ C2009-0-20099-7

[23] Guerrero MRB, Salinas Gutiérrez JM, Meléndez Zaragoza MJ, López Ortiz A, Collins-Martínez V. Optimal slow pyrolysis of apple pomace reaction conditions for the generation of a feedstock gas for hydrogen production. International Journal of Hydrogen Energy. 2016;**41**:23232-23237. DOI: 10.1016/j.ijhydene.2016.10.066

[24] Pratap A, Chouhan S. A slow pyrolysis of cotton stalk (*Gossypium arboretum*) waste for bio-oil production. Journal of Pharmaceutical, Chemical and Biological Sciences. 2015;**3**:143-149

[25] Demirbas MF, Balat M. Biomass pyrolysis for liquid fuels and chemicals: A review. Journal of Scientific and Industrial Research. 2007;**66**:797-804

[26] Polat S, Apaydin-Varol E, Pütün AE. Thermal decomposition behavior of tobacco stem. Part II: Kinetic analysis. Energy Sources, Part A: Recovery, Utilization, and Environmental Effects. 2016;**38**:3073-3080. DOI: 10.1080/ 15567036.2015.1129374

[27] Raveendran K, Ganesh A, Khilar KC. Pyrolysis characteristics of biomass and biomass components. Fuel. 1996;**75**: 987-998

[28] Greenhalf CE, Nowakowski DJ, Harms AB, Titiloye JO, Bridgwater AV. A comparative study of straw, perennial grasses and hardwoods in terms of fast pyrolysis products. Fuel. 2013;**108**: 216-230. DOI: 10.1016/j.fuel.2013. 01.075

[29] Vyazovkin S, Burnham AK, Criado JM, Pérez-Maqueda LA, Popescu C, Sbirrazzuoli N. ICTAC kinetics committee recommendations for performing kinetic computations on thermal analysis data. Thermochimica Acta. 2011;**520**:1-19. DOI: 10.1016/j. tca.2011.03.034

[30] Kirubakaran V, Sivaramakrishnan V, Nalini R, Sekar T, Premalatha M, Subramanian P. A review on gasification of biomass. Renewable and Sustainable Energy Reviews. 2009;**13**:179-186. DOI: 10.1016/j.rser.2007.07.001

[31] Williams PT, Besler S. The pyrolysis of rice husks in a thermogravimetric analyser and static batch reactor. Fuel. 1993;**72**:151-159. DOI: 10.1016/0016-2361(93)90391-E

[32] El-Sayed SA, Mostafa ME. Pyrolysis characteristics and kinetic parameters determination of biomass fuel powders by differential thermal gravimetric analysis (TGA/DTG). Energy Conversion and Management. 2014;**85**: 165-172. DOI: 10.1016/j.enconman.2014.05.068

[33] Mani T, Murugan P, Abedi J, Mahinpey N. Pyrolysis of wheat straw in a thermogravimetric analyzer: Effect of particle size and heating rate on devolatilization and estimation of global kinetics. Chemical Engineering Research and Design. 2010;**88**:952-958. DOI: 10.1016/j.cherd.2010.02.008

[34] Shen J, Igathinathane C, Yu M, Pothula AK. Biomass pyrolysis and combustion integral and differential reaction heats with temperatures using thermogravimetric analysis/differential scanning calorimetry. Bioresource Technology. 2015;**185**:89-98. DOI: 10.1016/j.biortech.2015.02.079

[35] Xiwen Y, Kaili X, Yu L. Assessing the effects of different process parameters on the pyrolysis Beha.: EBSCOhost. Bio Resources. 2017;**12**: 2748-2767. DOI: 10.1002/asi.22787

[36] Collazzo GC, Broetto CC, Perondi D, Junges J, Dettmer A, Dornelles Filho AA, et al. A detailed non-isothermal kinetic study of elephant grass pyrolysis from different models. Applied Thermal Engineering. 2017;**110**:1200-1211. DOI: 10.1016/j.applthermaleng.2016.09.012

[37] Ceylan S, Topçu Y. Pyrolysis kinetics of hazelnut husk using thermogravimetric analysis. Bioresource Technology. 2014;**156**:182-188. DOI: 10.1016/j.biortech.2014.01.040

[38] Haykiri-Acma H, Yaman S. Synergy in devolatilization characteristics of lignite and hazelnut shell during co-pyrolysis. Fuel. 2007;**86**:373-380. DOI: 10.1016/j.fuel.2006.07.005

[39] Wagenaar BM, Prins W, Swaaij WPM v. Flash pyrolysis kinetics of pine wood. Fuel Processing Technology. 1993;**36**:291-298

[40] Özsin G, Pütün AE. Co-pyrolytic behaviors of biomass and polystyrene: Kinetics, thermodynamics and evolved gas analysis. Korean Journal of Chemical Engineering. 2018;**35**:428-437. DOI: 10.1007/s11814-017-0308-6

[41] Gonzalez JF, Encinar JM, Canito JL, Sabio E, Chacon M. Pyrolysis of cherry stones: Energy uses of the different fractions and kinetic study. Journal of Analytical and Applied Pyrolysis. 2003; **67**:165-190

[42] Fiori L, Valbusa M, Lorenzi D, Fambri L. Modeling of the devolatilization kinetics during pyrolysis of grape residues. Bioresource Technology. 2012;**103**:389-397. DOI: 10.1016/j.biortech.2011.09.113

[43] Radmanesh R, Courbariaux Y, Chaouki J, Guy C. A unified lumped approach in kinetic modeling of biomass pyrolysis. Fuel. 2006;**85**:1211-1220. DOI: 10.1016/j.fuel.2005.11.021

[44] Cao L, Yuan X, Jiang L, Li C, Xiao Z, Huang Z, et al. Thermogravimetric characteristics and kinetics analysis of oil cake and torrefied biomass blends. Fuel. 2016;**175**:129-136. DOI: 10.1016/j.fuel.2016.01.089

[45] Tsai WT, Lee MK, Chang YM. Fast pyrolysis of rice husk: Product yields

and compositions. Bioresource Technology. 2007;**98**:22-28. DOI: 10.1016/j.biortech.2005.12.005

[46] Vieira FR, Luna CMR, Arce GLAF, Avila I. A study of biochar yield from slow pyrolysis of rice husk; In: 24th ABCM-Brazilian Association of Engineering and Mechanical Sciences; 2017. p. 1-6

[47] Ali N, Saleem M, Shahzad K, Chughtai A. Bio-oil production from fast pyrolysis of cotton stalk in fluidized bed reactor. Arabian Journal for Science and Engineering. 2015;**40**:3019-3027. DOI: 10.1007/s13369-015-1801-z

[48] Sidhu GK, Sandhya. Engineering properties of cotton stalks (*Gossypium hirsitum* L.), Indian. Journal of Agricultural Research. 2015;**49**:456-459. DOI: 10.18805/ijare.v49i5.5811

[49] Park J, Lee Y, Ryu C, Park Y. Slow pyrolysis of rice straw: Analysis of products properties, carbon and energy yields. Bioresource Technology. 2014; **155**:63-70. DOI: 10.1016/j.biortech.2013.12.084

[50] Song H, Andreas J, Minhou X. Kinetic study of Chinee biomass slow pyrolysis: Comparison of different kinetic models. Fuel. 2007;**86**:2778-2788. DOI: 10.1016/j.fuel.2007.02.031

[51] Mohanty P, Nanda S, Pant KK, Naik S, Kozinski JA, Dalai AK. Evaluation of the physiochemical development of biochars obtained from pyrolysis of wheat straw, timothy grass and pinewood: Effects of heating rate. Journal of Analytical and Applied Pyrolysis. 2013;**104**:485-493. DOI: 10.1016/j.jaap.2013.05.022

[52] Fu P, Li Z, Bai X, Altilia WY, Pietrangeli F, Torre AD, et al. Bio-oil production from fast pyrolysis of corn stalk in a fluidized bed. Applied Mechanics and Materials. 2014; **672–674**:143-146. DOI: 10.4028/www.scientific.net/AMM.672-674.143

[53] Cordella M, Berrueco C, Santarelli F, Paterson N, Kandiyoti R, Millan M. Yields and ageing of the liquids obtained by slow pyrolysis of sorghum, switchgrass and corn stalks. Journal of Analytical and Applied Pyrolysis. 2013; **104**:316-324. DOI: 10.1016/j.jaap.2013.07.001

[54] Zhang H, Xiao R, Wang D, He G, Shao S, Zhang J, et al. Biomass fast pyrolysis in a fluidized bed reactor under N_2, CO_2, CO, CH_4 and H_2 atmospheres. Bioresource Technology. 2011;**102**:4258-4264. DOI: 10.1016/j.biortech.2010.12.075

[55] Demirbas A. Effects of temperature and particle size on bio-char yield from pyrolysis of agricultural residues. Journal of Analytical and Applied Pyrolysis. 2004;**72**:243-248. DOI: 10.1016/j.jaap.2004.07.003

[56] Sousa J, Bezerra M, Almeida M, Moure G, Mesa-Perez J, Caramao E. Characteristics of bio-oil from the fast pyrolysis of elephant grass (*Pennisetum purpureum* schumach) in a fluidized bed reactor. American Chemical Science Journal. 2016;**14**:1-10. DOI: 10.9734/ACSJ/2016/25843

[57] Boateng AA, Mullen CA, Goldberg N, Hicks KB, Jung HJG, Lamb JFS. Production of bio-oil from alfalfa stems by fluidized-bed fast pyrolysis. Industrial and Engineering Chemistry Research. 2008;**47**:4115-4122. DOI: 10.1021/ie800096g

[58] Strezov V, Evans TJ, Hayman C. Conversion of elephant grass (*Pennisetum purpureum* schum) to bio-gas, bio-oil and charcoal. Bioresource Technology. 2008;**99**:8394-8399. DOI: 10.1016/j.biortech.2008.02.039

[59] Koçkar ÖM, Onay Ö, Pütün AE, Pütün E. Fixed-bed pyrolysis of

hazelnut shell: A study on mass transfer limitations on product yields and characterization of the pyrolysis oil. Energy Sources. 2000;**22**:913-924. DOI: 10.1080/00908310051128291

[60] Carrier M, Hugo T, Gorgens J, Knoetze H. Comparison of slow and vacuum pyrolysis of sugar cane bagasse. Journal of Analytical and Applied Pyrolysis. 2011;**90**:18-26. DOI: 10.1016/j.jaap.2010.10.001

[61] Chen T, Liu R, Scott NR. Characterization of energy carriers obtained from the pyrolysis of white ash, switchgrass and corn stover-biochar, syngas and bio-oil. Fuel Processing Technology. 2016;**142**:124-134. DOI: 10.1016/j.fuproc.2015.09.034

[62] Onay Ö, Beis SH, Koçkar ÖM. Pyrolysis of walnut shell in a well-swept fixed-bed reactor. Energy Sources. 2004;**26**:771-782. DOI: 10.1080/00908310490451402

[63] Yuan HR, Liu RH. Study on pyrolysis kinetics of walnut shell. Journal of Thermal Analysis and Calorimetry. 2007;**89**:983-986. DOI: 10.1504/IJGEI.2008.018006

[64] Suttibak S, Sriprateep K, Pattiya A. Production of bio-oil from pine sawdust by rapid pyrolysis in a fluidized-bed reactor. Energy Sources, Part A: Recovery, Utilization, and Environmental Effects. 2015;**37**:1440-1446. DOI: 10.1080/15567036.2011.631091

[65] Uzun BB, Kanmaz G. Effect of operating parameters on bio-fuel production from waste furniture sawdust. Waste Management & Research. 2013;**31**:361-367. DOI: 10.1177/0734242X12470402

[66] Wang Z, Cao J, Wang J. Pyrolytic characteristics of pine wood in a slowly heating and gas sweeping fixed-bed reactor. Journal of Analytical and Applied Pyrolysis. 2009;**84**:179-184. DOI: 10.1016/j.jaap.2009.02.001

[67] Sundaram EG, Natarajan E. Pyrolysis of coconut shell: An experimental investigation. Journal of Engineering Research. 2009;**6**:33. DOI: 10.24200/tjer.vol6iss2pp33-39

[68] Mythili R, Venkatachalam P, Subramanian P, Uma D. Characterization of bioresidues for biooil production through pyrolysis. Bioresource Technology. 2013;**138**:71-78. DOI: 10.1016/j.biortech.2013.03.161

[69] Mythili R, Subramanian P, Uma D. Biofuel production from *Prosopis juliflora* in fluidized bed reactor. Energy Sources, Part A: Recovery, Utilization, and Environmental Effects. 2017;**39**:741-746. DOI: 10.1080/15567036.2016.1261205

[70] Chandrasekaran A, Ramachandran S, Subbiah S. Modeling, experimental validation and optimization of *Prosopis juliflora* fuelwood pyrolysis in fixed-bed tubular reactor. Bioresource Technology. 2018;**264**:66-77. DOI: 10.1016/j.biortech.2018.05.013

[71] Xu J, Qiao L. Mathematical modeling of coal gasification processes in a well-stirred reactor: Effects of devolatilization and moisture content. Energy & Fuels. 2012;**26**:5759-5768. DOI: 10.1021/ef3008745

[72] Narataruksa P, Tungkamani S, Pana-Suppamassadu K, Keeratiwintakorn P, Nivitchanyong S, Hunpinyo P, et al. Conversion enhancement of tubular fixed-bed reactor for Fischer-Tropsch synthesis using static mixer. Journal of Natural Gas Chemistry. 2012;**21**:435-444. DOI: 10.1016/S1003-9953(11)60388-5

[73] Wu J, Zhang H, Ying W, Fang D. Simulation and analysis of a tubular fixed-bed Fischer-Tropsch synthesis

reactor with co-based catalyst. Chemical Engineering and Technology. 2010;**33**: 1083-1092. DOI: 10.1002/ceat. 200900610

[74] Kunii D, Levenspiel O. Fluidization Engineering. 2nd ed. Boston: Elsevier; 1991

[75] Ciesielczyk W, Iwanowski J. Analysis of fluidized bed drying kinetics on the basis of interphase mass transfer coefficient. Journal of Drying Technology. 2006;**24**:1153-1157

[76] Agarwal PK, Genetti WE, Lee YY. Drying and devolatilization of Mississippi lignite in a fluidized-bed. American Chemical Journal. 1984:**29**

Chapter 5

Waste Plastics Valorization by Fast Pyrolysis and in Line Catalytic Steam Reforming for Hydrogen Production

Itsaso Barbarias, Aitor Arregi, Maite Artetxe, Laura Santamaria, Gartzen Lopez, María Cortazar, Maider Amutio, Javier Bilbao and Martin Olazar

Abstract

This chapter summarizes the most recent results obtained in the plastic waste pyrolysis-reforming strategy for hydrogen production. An original two-reactor configuration consisting of a conical spouted bed reactor for the pyrolysis step and a fluidized bed reactor for the pyrolysis volatile reforming is proposed. The fundamental aspects and challenges of this joint process are discussed in detail, and the prospects for the full-scale implementation of this valorization route are assessed. Thus, the influence the main reforming parameters (temperature, space time and steam/plastic ratio) have in the pyrolysis-reforming of HDPE on product yields and catalyst stability are reported. Moreover, the role played by plastic composition on process performance is also described by studying the influence of following polymers: high density polyethylene (HDPE), polypropylene (PP), polyethylene terephthalate (PET) and polystyrene (PS). The operating conditions used for the valorization of different plastics have been as follows: pyrolysis temperature of 500°C, reforming temperature of 700°C, space time of 16.7 $g_{catalyst}$ min $g_{plastic}^{-1}$ and steam/plastic ratio of 4.

Keywords: hydrogen, pyrolysis, reforming, waste plastics, spouted bed

1. Introduction

The current overall hydrogen production has reached a value of 7.7 EJ/year, and an increase to 10 EJ/year is expected by the year 2050 [1]. Hydrogen demand is mainly related to ammonia production (51%), oil refining (31%), methanol production (10%), and other uses (8%). Moreover, new fossil sources of lower quality than conventional oil are increasingly been used, but they require severe hydroprocessing for their upgrading. This is the case of oil sands, which surely will contribute to meeting the increasing requirements of hydrogen.

Approximately 96% of the world hydrogen produced is of fossil origin. Thus, 48% is produced from natural gas reforming, 30% from oil fractions and the remaining 18% by means of coal gasification [2]. Accordingly, only 4% is produced from renewable sources, such as water electrolysis.

The future of the hydrogen market is conditioned by the need to reduce CO_2 emissions, which requires obtaining it from raw materials and renewable energy sources. The so-called hydrogen economy contemplates a future scenario in which this gas produced in a clean and economical way from a wide range of renewable sources would contribute to meeting the society's energy needs through combustion engines or preferably, by means of fuel cells, to a more efficient energy generation. This proposal would help to decentralize and diversify the energy system and reduce the current dependence on fossil fuels and air pollution [3].

In this scenario, the processes aimed at hydrogen production from biomass and waste are gaining growing attention [1, 4–6], with the thermochemical routes being those with best perspectives for their full scale implementation. Thus, hydrogen production from waste plastics has been mainly approached by means of steam gasification and pyrolysis-reforming processes [6].

Gasification allows for the conversion of waste plastics into a gaseous stream with varying contents of H_2, CO, CO_2, CH_4 and N_2 depending on process conditions and the gasifying agent used. The main advantage of gasification in relation to other thermochemical processes lies in its flexibility to valorize plastics of different composition or mixtures or plastics with other feedstocks. The main technologies for waste plastics gasification are those previously developed for biomass and coal gasification. It is to note that fluidized bed reactors have been widely used in the gasification of plastics [7–13]. However, these studies were mainly performed using air as fluidizing agent, so that the resulting gas has low hydrogen content and its main interest is restricted to energy production. The dual fluidized bed technology has been successfully applied to the steam gasification of plastics [13] and plastics mixtures with coal and biomass [14, 15]. The relevance of this technology is related with the fact that it allows overcoming the problems associated by the endothermic nature of the process. In spite of the promising results reported by Wilk et al. [13] in the gasification of different waste plastics, the high tar content in the syngas (above 100 g m^{-3}) is a great challenge. However, waste plastic gasification has been scarcely studied in fixed bed reactors. Moreover, the results obtained are of preliminary nature as the reactors used are of laboratory scale and operate in batch regime [16–19]. Erkiaga et al. [20, 21] used the conical spouted bed reactor in the steam gasification of HDPE, and a hydrogen rich syngas was obtained with moderate tar contents. Moreover, this reactor design was also used subsequently in the co-gasification of HDPE and biomass (pine sawdust) with promising results [22]. The high temperatures attained in plasma reactors are also suitable for tar elimination, and this technology has therefore been proposed for the gasification of waste plastics [23, 24].

The hydrogen production from plastics via pyrolysis-reforming strategy has been recently proposed as an alternative to conventional gasification. The interest of this strategy lies in the fact that a tar free syngas is produced due to the high activity of the reforming catalyst. Moreover, the mild conditions in the reforming step and the physical separation of waste plastic impurities and the catalysts improve catalyst stability. Nevertheless, this strategy has been scarcely studied and the results reported were mainly obtained in lab scale units [6]. The studies by the team of Professor Williams [25–28] were carried out in a system in which pyrolysis is carried out in discontinuous regime in a fixed bed reactor and the volatiles are reformed in a second fixed bed. The reaction equipment used by the team of Professor Yoshikawa [29, 30] is a step forward, as it operates in continuous regime, although the use of fixed beds for pyrolysis and reforming can involve serious limitations for the process scale up. Czernik and French [31] made a step forward in this technology by studying the continuous pyrolysis and in-line reforming of polypropylene in a system of two in-line fluidized beds, with a commercial naphtha reforming catalyst. Erkiaga et al. [32] originally proposed a two-step reactor unit operating in continuous regime by combining a spouted bed reactor for the plastic fast pyrolysis and a fixed bed for the catalytic steam reforming. However, the fixed bed reactor led to an excessive coke formation and severe operational problems subsequent to some minutes operation. Accordingly, the fixed bed reactor was replaced by a fluidized bed reactor in subsequent studies to overcome the mentioned problems [33–36].

This chapter describes the main results obtained in the pyrolysis and in-line reforming of plastics in a system composed of a spouted bed for the pyrolysis step and a fluidized bed for the reforming one. Section 2 provides background information on the application of conical spouted beds to plastics pyrolysis. In addition, the results obtained in the pyrolysis of different plastics are also reported in this section, as long as the pyrolysis step product distribution is critical for the evaluation of the subsequent reforming step. Finally, Section 3 deals with the effect operating conditions (temperature, space time and S/P ratio) have on the reforming step performance. Moreover, the influence the polymer type in the feed to the pyrolysis step has on reforming step yields has also been evaluated. Thus, the interest of this chapter is centered on the maximization of hydrogen production under different conditions by treating plastics of different composition. Furthermore, one of the main challenges of this process, as is catalyst stability, has also been approached in detail.

2. Waste plastics pyrolysis in conical spouted beds

Pyrolysis (or thermal cracking) for the valorization of waste plastic is currently receiving great attention, especially for addition polymers, which accounts for most plastics in the municipal solid waste stream. One of the main advantages of pyrolysis lies in its efficiency for the recovery of monomers and for the production of automotive and hydrogen fuels, with reduced emission of pollutants [37–40].

The pyrolysis process is flexible and may treat plastic mixtures in the municipal solid waste, and those from automobile recycling, electronics and computers or construction, as well as mixtures of these materials with others (such as biomass) [41]. The requirements for scaling up and operation in continuous regime have led to the proposal of a significant number of technologies provided with very different reactors, such as fluidized beds, screw kilns, stirred or microwave reactors [37, 38]. Amongst these technologies, the fluidized bed reactor has been widely used because it ensures good heat and mass transfer, and therefore isothermal operation [42]. The main international reference in the development of this technology is known as the Hamburg process [43–45].

One of the main challenges in the pyrolysis of plastics is the sticky behavior of fused plastic particles, which causes agglomeration and subsequent defluidization in fluidized bed reactors [46]. The high heat transfer rates and vigorous cyclic movement of the particles in the conical spouted bed reactor, as well as the role of the spout to break up the agglomerates, are suitable features for continuous pyrolysis of plastics. Therefore, this technology allows operating below 500°C without defluidization problems [47]. Accordingly, the conical spouted bed has been successfully applied to the continuous pyrolysis of different polymeric materials [48–52]. Moreover, this technology has also been applied to the pyrolysis of biomass and, in fact, a 25 kg h^{-1} pilot plant has been developed for this process [53, 54].

The main interest in analyzing the results obtained in the pyrolysis of the plastics lies in the fact that it constitutes the first step of the integrated pyrolysis-reforming strategy, and therefore the volatile products of the pyrolysis process are the reactants of the catalytic reforming step. Consequently, it is essential to know the composition of the volatile stream leaving the pyrolysis step. Accordingly, the composition of the volatile stream for the pyrolysis of polyolefins (HDPE and PP), PET and PS, are analyzed in this section.

The different nature of the plastics studied led to volatile streams of different composition in their pyrolysis. It is to note that HDPE and PP are made up of saturated hydrocarbon chains, the thermal degradation of these chains mainly occurs by random radical scission mechanism, which gives way to a wide product distribution ranging from light gases to waxes. The fast pyrolysis conditions of the conical spouted bed reactor and especially the short residence time attained favor the selective formation of primary products, i.e., waxes and hydrocarbons in the diesel range, with the yield of gases, gasoline and aromatics being very low. However, in the PS chains aromatic rings are included, thus, the depolymerisation of this plastic leads to the formation of monomers (styrene), oligomers and other secondary products of aromatic nature. Due to the low temperature and the suitable features of the reactor used in this study, the product stream obtained in PS pyrolysis is mainly made up of styrene. Finally, PET is produced by polymerization of terephthalic acid and ethylene Glycerol. Accordingly, a wide product distribution with oxygenates and aromatics are formed in PET pyrolysis. The different nature of the pyrolysis product stream has a remarkable influence on its behavior in the reforming step, and the following section deals with this aspect.

According to the results obtained in previous studies, pyrolysis temperature was fixed at 500°C [47, 49, 52]. This temperature ensures a stable process without operational problems and minimizes secondary reactions of pyrolysis products and energy requirements. Moreover, below this temperature the plastic pyrolysis kinetic is very slow. Although the polymer degradation in the pyrolysis-reforming process occurs under steam atmosphere, the results obtained in previous studies using steam are very similar to those obtained using nitrogen as fluidizing agent, and therefore steam has hardly any effect at this temperature [33, 35]. This result is associated with the mild pyrolysis conditions used, 500°C, which hinders steam reactivity. The experiments were performed with continuous plastic feed and using a bed of inert sand in the pyrolysis of all the polymers studied.

The results obtained in the continuous pyrolysis of different plastics, HDPE, PP, PS and PET at 500°C are summarized in **Figure 1**. Due to the similar nature of HDPE and PP, similar product yields are obtained in their pyrolysis. As observed in **Figure 1**, the pyrolysis of polyolefins (HDPE and PP) leads to high yields of waxes (C_{21+}) and liquid hydrocarbons in the diesel range (C_{12}—C_{20}), with the yield of the gasoline fraction being low and that of gases below 1.5 wt%. It is to note that the solid residue formed in the pyrolysis of polyolefins is negligible, and therefore the whole product stream is fed into the fluidized bed for its reforming. Waxes are made up of long chain hydrocarbons and are solid at room temperature. However, these products do not cause any operational problem in the pyrolysis-reforming process as long as they are fully converted into gaseous products in the reforming step.

On the other hand, PET showed a wider product distribution that those observed in the case of polyolefins. A remarkable solid residue yield (7 wt%) was obtained in the pyrolysis of PET, with this residue being obtained coating the sand particles. The main product fraction in PET pyrolysis is the gaseous stream, whose yield is of 42.8 wt%. This gas is mainly made up of carbon monoxide and dioxide. In addition, a remarkable yield of a solid fraction (benzoic and benzoylformic acids) is obtained at room temperature, which accounts for 37.4 wt%. However, the yield of oil is the lowest (12.8 wt%) in the polymers studied.

The pyrolysis of PS leads to a high monomer recovery, with the yield of styrene being 70.6 wt%. The gaseous product yield is very low, 1.5 wt%, and that of the solid residue negligible. Thus, the product distribution obtained in PS pyrolysis is made up of styrene and other secondary products of aromatic nature.

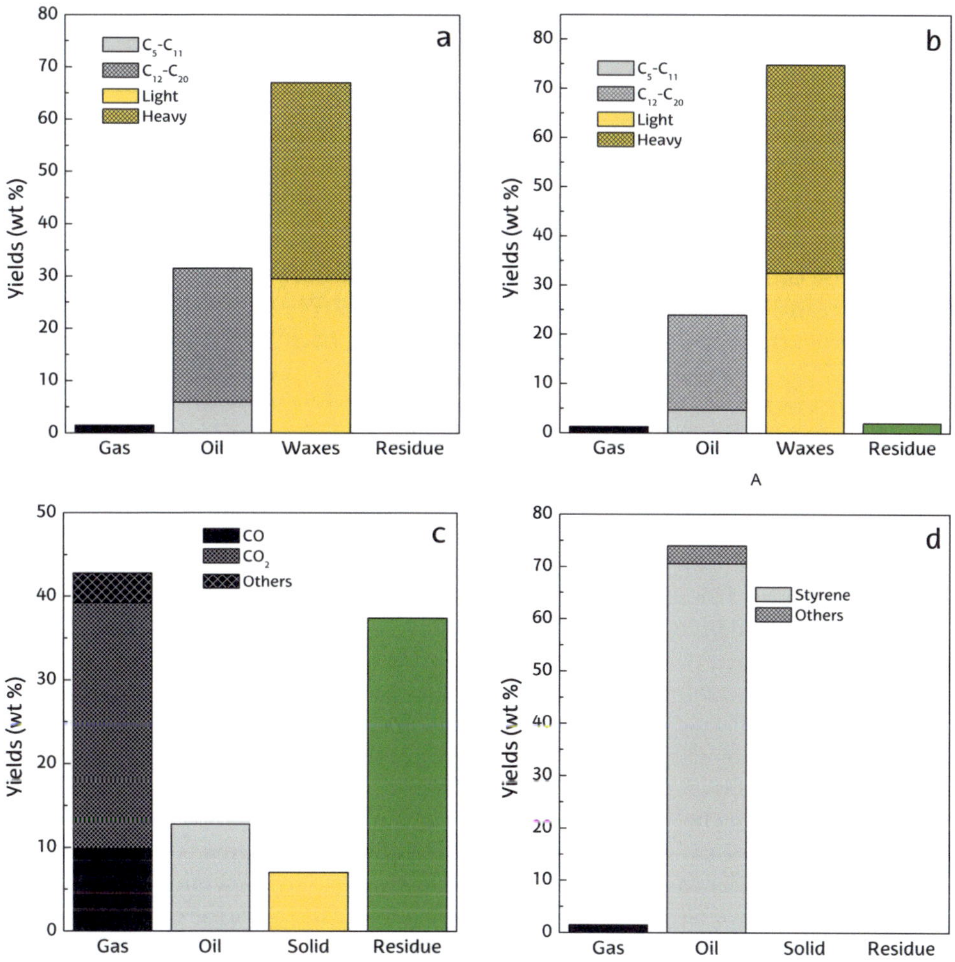

Figure 1.
Product distributions obtained in the continuous pyrolysis of HDPE (a), PP (b), PET (c) and PS (d) in a conical spouted bed reactor at 500°C.

3. Pyrolysis and in-line reforming of plastics

The fast pyrolysis of waste plastics has been extensively studied in the literature [38, 55, 56], due to its interest for producing fuels and chemicals. Recently, the pyrolysis and in-line reforming route for hydrogen production has gained attention, with several studies published over the last years [30, 31, 57]. In this scenario, the optimization of the process operating conditions is essential for its industrial implementation.

Thus, the following section shows the results obtained in the reforming of HDPE derived pyrolysis volatiles carried out using the configuration described above, i.e., the conical spouted bed reactor for the pyrolysis step and the fluidized bed reactor for the catalytic steam reforming step. The aim of this research was to study how different operating conditions affect the reforming of HDPE pyrolysis volatiles. The parametric study includes the effect of temperature, space time and steam/plastic (S/P) ratio on the conversion, hydrogen production and the yields of the products both at zero time on stream and over time, taking into account the evolution of these results as a consequence of the catalyst deactivation.

3.1 Parametric study of HDPE pyrolysis and in-line reforming

The results obtained in the parametric study carried out at zero time on stream are described below.

3.1.1 Effect of temperature

This section deals with the effect the reforming temperature in the 600–700°C range has on conversion, hydrogen production and gaseous products yields at zero time on stream. The operating conditions selected to carry out this study have been the following: (i) temperature in the pyrolysis step has been 500°C; (ii) a high space time of 16.7 $g_{catalyst}$ min g_{HDPE}^{-1} (corresponding to 12.5 g of catalyst); (iii) steam/plastic (S/P) ratio of 4. These conditions have been established in order to avoid operational problems. The space time value has been given as the amount of catalyst used by mass unit of the plastic fed into the pyrolysis reactor. High value of space time (16.7 $g_{catalyst}$ min g_{HDPE}^{-1}) has been used in order to ensure high initial conversion of the volatiles coming from pyrolysis step and ensure a slow deactivation of the catalyst. Moreover, using a S/P ratio of 4 catalyst stability is guaranteed, given that coke gasification is favored and at the same time hydrogen production is increased.

The gaseous products quantified are as follows: H_2, CO_2, CO, CH_4 and light hydrocarbon fraction (C_2—C_4), composed mainly of ethylene, ethane, propylene and propane. The unreformed liquid fraction in the volatile stream coming from pyrolysis step has been called C_{5+} fraction. This fraction is conventionally called tar and should be removed in order for the product gas to be used as synthesis gas or conduct a further separation to isolate hydrogen.

The evolution of the conversion and hydrogen production with temperature is shown in **Figure 2a** and **b** shows the evolution of gaseous products yields with temperature. In order to quantify the results obtained in the reforming step, some reaction indices have been defined. The conversion is defined as the carbon conversion efficiency, i.e., the percentage of carbon moles contained in the plastic feed that are recovered in the gaseous products. **Figure 2a** shows that the conversion is greater than 94% in the temperature range studied. In addition, an increase in temperature enhances the reforming reaction, achieving a conversion of 98% at 700°C. Therefore, at this temperature only 2% of the carbon fed, which is in the C_{5+} fraction of the product stream, remains without transforming into gas. It should be noted that, at the temperature range studied, the conversion of hydrocarbons entering the reforming reactor (mostly waxes as aforementioned in the previous section) is almost full. This trend is characteristic of steam reforming reactions [29, 58], which is because the endothermic hydrocarbon reforming reactions and cracking reactions are enhanced as temperature is increased.

Another interesting index for quantify the results obtained is hydrogen production. This index is calculated as the mass of hydrogen produced per plastic mass unit fed into the pyrolysis step. **Figure 2a** shows the evolution of hydrogen production with temperature. The production of hydrogen increases with temperature, reaching a value of 37.3 wt% at 700°C. In fact, as temperature is increased the endothermic reaction of reforming is favored, enhancing hydrogen production. Based on the mass balance of hydrogen, approximately 63% of this hydrogen is formed from steam and the remaining 37% from the plastic. This hydrogen production is slightly higher than that obtained by Czernik and French [31], 34 wt%, in their study of pyrolysis-reforming of PP.

The increase in conversion and hydrogen production with temperature in the catalytic steam reforming of municipal solid wastes [59] is explained because all the endothermic steps involved are favored: (i) initial pyrolysis of waste; (ii) char

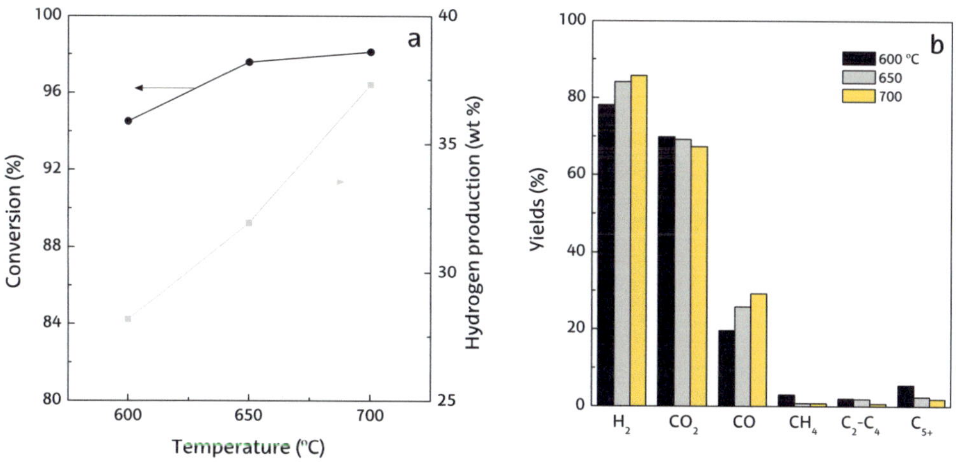

Figure 2.
(a) Evolution of HDPE conversion and hydrogen production with temperature; (b) effect of reforming temperature on product yields. Reaction conditions: $T_{pyrolysis}$, 500°C; space time, 16.7 $g_{catalyst}$ min g_{HDPE}^{-1} and S/P ratio: 4.

gasification reactions; (iii) cracking and steam reforming of the tar. In the latter case, the increase in temperature enhances the endothermic reactions of reforming, as well as cracking.

Figure 2b shows the effect reforming temperature in the 600–700°C range has on products yields. Product yields are the reaction indices defined for the assessment of the reforming step performance. The yield of C containing individual compounds is defined by mass unit of plastic in the feed, and the hydrogen yield is a percentage of the maximum allowed by stoichiometry, which accounts for the hydrogen coming from the polymer and the steam. **Figure 2b** reveals that an increase in the reforming temperature leads to an increase in the yields of hydrogen, from 78.1% at 600°C to 85.7% at 700°C. Simultaneously, it is observed that temperature increase favors the formation of CO, obtaining a yield of 29.2% at 700°C. The effect of temperature on the CO_2 yield is lower, since it remains constant due to the shift in the equilibrium of the water-gas shift (WGS) exothermic reaction.

The increase in the yield of hydrogen and CO is because an increase in the reforming temperature favors the reforming of waxes, light hydrocarbons and CH_4. Thus, although CH_4, C_2—C_4 and C_{5+} yields are very low in the whole temperature range studied, a slight decrease is observed as temperature is increased, obtaining yields of 0.8, 0.7 and 1.9% at 700°C, respectively. This result indicates the reforming of these compounds is enhanced selectively over their formation by cracking. On the other hand, it must be taken into account that the WGS reaction is exothermic, and therefore its thermodynamic equilibrium shift towards CO formation by increasing temperature.

3.1.2 Effect of space time

The effect the space time has on the reforming step has been studied in the 2.8–20.8 $g_{catalyst}$ min g_{HDPE}^{-1} range, by varying the amount of the catalyst, with the plastic feed rate being 0.75 g min^{-1}. The remaining conditions are as follows: (i) temperature in the pyrolysis step, 500°C; (ii) temperature in the reforming step, 700°C; (iii) S/P ratio, 4. As in the previous section, the effect of space time on the conversion, hydrogen production and gaseous products yields has been studied.

Figure 3 shows the evolution with space time of conversion and hydrogen production (**Figure 3a**) and of the products yields (**Figure 3b**). As observed in **Figure 3a**,

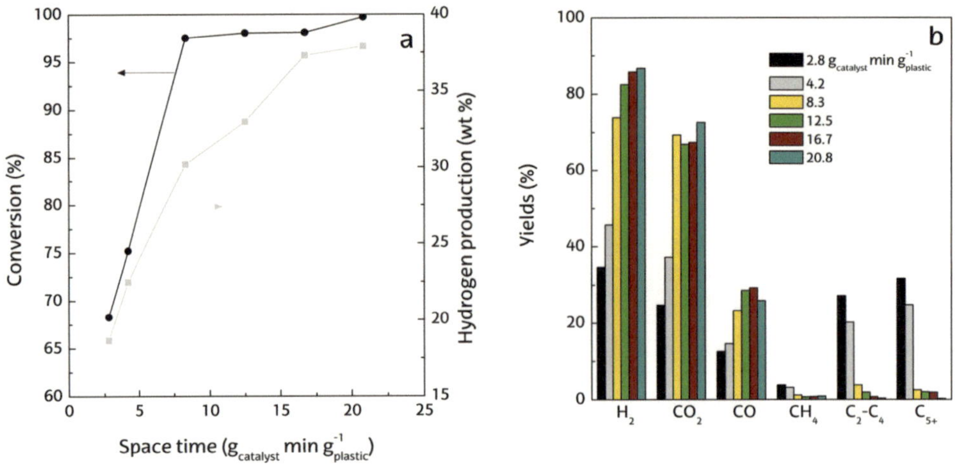

Figure 3.
(a) Evolution of HDPE conversion and hydrogen production with space time; (b) effect of space time on product yields. Reaction conditions: $T_{pyrolysis}$, 500°C; $T_{reforming}$, 700°C; and S/P ratio, 4.

there is a sharp increase in conversion when space time is increased from 2.8 (68.3%) to 8.3 $g_{catalyst}$ min g_{HDPE}^{-1} (97.5%). An increase in space time above the latter value only contributes to slightly increasing conversion, which is almost full (99.7%) for a space time of 20.8 $g_{catalyst}$ min g_{HDPE}^{-1}. Regarding hydrogen production, it increases from 18.6 wt% at 2.8 to 38 wt% at 20.8 $g_{catalyst}$ min g_{HDPE}^{-1}. The production of hydrogen increases more significantly in the space time range from 2.8 to 8.3 $g_{catalyst}$ min g_{HDPE}^{-1}, whereas in the 8.3–20.8 $g_{catalyst}$ min g_{HDPE}^{-1} range the increase is milder.

Figure 3b shows the effect space time in the 2.8–20.8 $g_{catalyst}$ min g_{HDPE}^{-1} range has on product yields. By enhancing the extent of reforming reactions and WGS reactions, the increase in space time increases the yields of hydrogen, CO_2 and CO, whereas it decreases the yields of CH_4, C_2—C_4 and C_{5+}. An increase in space time (**Figure 3a**) from 2.8 to 8.3 $g_{catalyst}$ min g_{HDPE}^{-1} leads to a pronounced increase in the yield of hydrogen and CO_2 to 73.8 and 69.3%, respectively. In the mentioned range of space time, the increase in CO yield is less significant, reaching a value of 23.2% for 8.3 $g_{catalyst}$ min g_{HDPE}^{-1}. Therefore, it seems that increasing space time in this range the WGS reaction is selectively favored. It is also important to highlight the decrease in the yields of CH_4, C_2—C_4 and C_{5+}, especially in the case of C_2—C_4 and C_{5+}, whose yields decrease from 27.2 to 3.8%, and from 31.7 to 2.4%, respectively, in the 2.8–8.3 $g_{catalyst}$ min g_{HDPE}^{-1} range.

For space times higher than 8.3 $g_{catalyst}$ min g_{HDPE}^{-1} there is hardly any change in the yields. However, in the case of hydrogen, its yield continues increasing, reaching 86.7% for a space time of 20.8 $g_{catalyst}$ min g_{HDPE}^{-1}. On the other hand, an increase in space time from 8.3 to 16.7 $g_{catalyst}$ min g_{HDPE}^{-1} hardly affects CO_2 yield, whereas it leads to a slight decrease in CO, which reaches a maximum value of 29.3%.

The remarkable effect of space time on conversion and gaseous product yields is well established in the literature in both the pyrolysis and reforming of plastics [60] and the gasification and reforming of municipal solid waste [59, 61, 62], which is explained by the enhancement of the reforming and WGS reactions.

3.1.3 Effect of steam/plastic (S/P) ratio

In this section, the effect S/P ratio in the 3–5 range has on conversion, hydrogen production and products yields at zero time on stream is shown. The operating conditions selected to carry out this study have been the following: (i) temperature of the pyrolysis step, 500°C; (ii) temperature of the reforming step, 700°C; (iii) space time, 16.7 $g_{catalyst}$ min g_{HDPE}^{-1}. The effect of S/P ratio is directly related to the steam partial pressure, thus the steam partial pressure in the reaction medium is more significant when this ratio is increased.

Figure 4 shows the evolution of conversion and hydrogen production with the steam/plastic (S/P) ratio (**Figure 4a**), as well as the evolution of products yields with S/P ratio (**Figure 4b**). **Figure 4a** shows that in the S/P range studied conversion is over 98%, reaching almost 99.6% with a ratio of 5. By increasing the S/P ratio from 3 to 4, the effect of the ratio is not very relevant and, as mentioned above, by increasing to 5 conversion is almost full. The reaction rate of the reforming and WGS reactions are enhanced by increasing the steam partial pressure in the reaction environment, which explains the high conversions obtained. Consequently, hydrogen production significantly increases in the S/P range studied, from 31.0 wt% for a S/P ratio of 3 to 38.1 wt% for a S/P ratio of 5.

Figure 4b displays the effect S/P ratio in the 3–5 range has on product yields. It is observed that an increase in this ratio leads to a progressive increase in the hydrogen and CO_2 yields and a decrease in those of CO, CH_4, C_2—C_4 and C_{5+}. This effect is due to the higher concentration of steam in the reaction environment, which favors the displacement of both hydrocarbons reforming reactions and the WGS reaction. By favoring all the reactions involved in the reforming, an increase in the S/P ratio enhances the yield of hydrogen and CO_2, up to 92.5 and 80.9% for a ratio of 5. Consequently, the yield of CO decreases to 18.1%. The yields of CH_4, C_2—C_4 and C_{5+}, which are small for a S/P ratio of 4, decrease until becoming insignificant.

As observed, an increase in S/P ratio causes a positive effect on conversion and hydrogen production; however, an excessive increase affects significantly on the energy efficiency of the process.

Qualitatively similar effects of S/P ratio were observed in the pyrolysis-reforming of PP [63, 64], LDPE and PS [63] and municipal plastic wastes [65], which is a widely proven fact in the gasification of different polymeric materials [58, 66].

3.1.4 Catalyst deactivation

Knowledge of catalyst deactivation is essential for its choice, design of the reactor and establishment of the optimal strategy for the operation of industrial

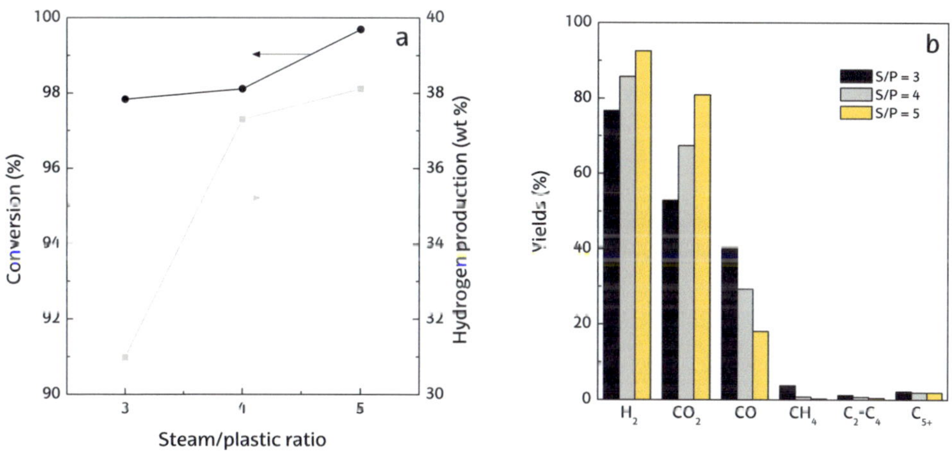

Figure 4.
(a) Evolution of HDPE conversion and hydrogen production with S/P ratio; (b) effect of the S/P ratio on product yields. Reaction conditions: $T_{pyrolysis}$, 500°C; $T_{reforming}$, 700°C; and space time, 16.7 $g_{catalyst}$ min g_{HDPE}^{-1}.

catalytic processes. It is well documented that coke deposition is the main cause of catalyst deactivation in the pyrolysis-reforming of plastics [63, 67], as well as in the reforming of hydrocarbons and oxygenates (methanol, DME, ethanol, bio-oil) [68, 69]. This deposition is especially important for Ni based catalysts.

The commercial catalyst used in this study does not undergo a significant Ni sintering because it is designed to use in CH_4 reforming, which is a reaction carried out at a temperature of around 850°C, i.e., higher than the highest used in this study (700°C). The plastics used do not contain S, and therefore the deactivation cause is the deposition of coke. Three different types of coke have been identified in the reforming of CH_4 on Ni catalyst: pyrolytic, encapsulating and filamentous [70]. Numerous studies support this classification of the coke deposited in the reforming of CH_4, hydrocarbons and biomass derivates.

Wu and Williams [26] have verified by transmission electron microscopy (TEM) that the coke deposited on Ni catalyst used in the production of hydrogen by pyrolysis-reforming of PP is composed of two types of coke, amorphous and filamentous, with the latter having similar characteristics as carbon nanotubes and graphene (according to their graphitization degree). Considering the commercial interest of filamentous coke for these applications, Acomb et al. [63] have studied the simultaneous production of carbon nanotubes and hydrogen through pyrolysis-reforming of PP, which improves the economy of the process and provides an additional incentive for the viability of plastic recycling.

In this section, a study is carried out of the effect catalyst deactivation has on conversion and product yields with time on stream. The reaction conditions used are as follows: (i) temperature in the pyrolysis step, 500°C; (ii) temperature in the reforming step, 700°C; (iii) space time, 16.7 $g_{catalyst}$ min g_{HDPE}^{-1} and (iv) S/P ratio, 4. The study has been carried out under conditions in which a slow deactivation of the catalyst is observed, evaluating the effect of the incipient deposition of coke and the evolution of this deposition up to 123 min of reaction.

Figure 5 shows the effect catalyst deactivation has on plastics conversion (**Figure 5a**) and product yields (**Figure 5b**). As observed, whilst the reforming reaction progresses with time on stream the conversion of plastics decreases due to the deactivation undergone by the catalyst, decreasing linearly from 98.1% at zero time on stream to 90.3% in 123 min. Besides, comparing the evolution of conversion (**Figure 5a**) and the evolution of product yields (**Figure 5b**), the decrease in H_2 and CO_2 yields decrease in a similar trend to that of conversion, as catalyst deactivation hinders the reforming reaction, as well as the WGS reaction. Consequently, CH_4, C_2—C_4 and C_{5+} hydrocarbon yields increase. As observed in **Figure 5**, the conversion and the yields of H_2 and CO_2 decrease linearly and those of CH_4, C_2—C_4 and C_{5+} hydrocarbons increase linearly. The lineal catalyst deactivation rate observed should be attributed to: (i) high temperature favoring the kinetics of reforming reactions due to their endothermic nature, which leads to a low concentration of hydrocarbons in the reaction medium, and therefore low coke deposition rate (low catalyst deactivation); (ii) high temperature enhances the gasification of the deposited coke and so minimize catalyst deactivation [68].

As a result of reforming activity decrease, CH_4, C_2—C_4 and C_{5+} fraction yields increase over time, which is more noticeable as the molecular weight of the hydrocarbons increases. This trend is explained by the reforming of C_{5+} hydrocarbons in parallel with the thermal cracking of these compounds towards CH_4 and C_2—C_4. Consequently, as reforming rate decreases due to catalyst deactivation the thermal cracking reactions are favored. This cracking phenomenon is well known in deactivated catalysts and is more important as the molecular weight of aliphatic compounds is higher [71].

It is established in the literature that paraffin capacity for coke formation is greater as their molecular weight is higher [72]. Similarly, the capacity for coke

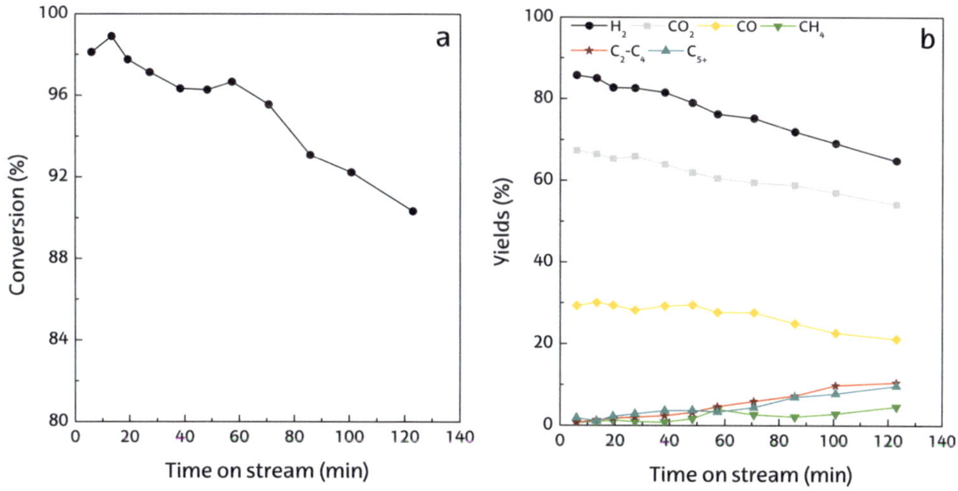

Figure 5.
(a) Evolution of HDPE conversion with time on stream; (b) evolution of product yields with time on stream. Reaction conditions: $T_{pyrolysis}$*, 500°C;* $T_{reforming}$*, 700°C; space time, 16.7* $g_{catalyst}\ min\ g_{HDPE}^{-1}$*; S/P = 4.*

formation of ethylene and benzene, is higher than that of paraffin compounds [71]. It is also well recognized that the coke formed on a Ni catalyst at high temperature is more filamentous and with less capacity for blocking Ni sites [68, 71].

Figure 6 displays a temperature programmed oxidation (TPO) profile of the coke deposited on the catalyst after 123 min reaction. Moreover, an average coke deposition rate has been quantified in order to calculate the extent of coke deposition. According to the TPO profile obtained, it has been determined that the average coke deposition rate is 0.7 $mg_{coke}\ g_{catalyst}^{-1}\ g_{HDPE}^{-1}$. Furthermore, two different peaks may be observed in **Figure 6**, which are evidence of two cokes of different

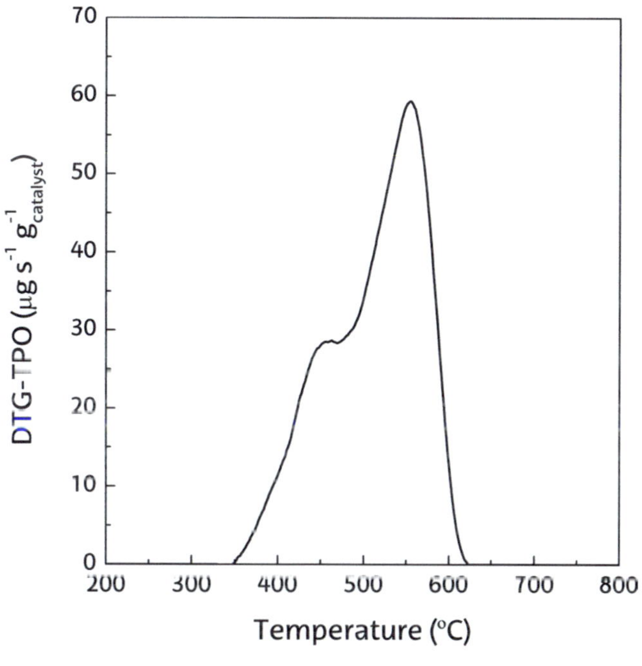

Figure 6.
TPO profile of the coke deposited on the catalyst.

nature and location on the catalyst. The main peak appears at 550°C and accounts for a partially structured coke, whereas the second peak at 430°C corresponds to an encapsulating coke, which is deposited on the Ni metal.

The encapsulating coke is deposited close to the actives sites, which promote coke combustion. Furthermore, this amorphous structure hinders the adsorption of reactants, which increases the concentration of non-converted hydrocarbons in the reaction environment, and consequently deactivates the catalyst. On the other hand, the combustion of the partially structured coke occurs at higher temperature due to the further location of the coke from the metallic sites, thereby hindering its combustion [63, 73]. This type of deposited coke corresponds to the filamentous coke, which generally does not affect directly to catalyst deactivation, as it does not block the active sites [69]. However, when its content is severe, it blocks the active sites, hindering the adsorption of reactants, and consequently increasing the concentration of non-converted hydrocarbons in the reaction medium.

The aforementioned identification of two types of coke has been corroborated based on the scanning electron microscopy (SEM) image of the deactivated catalyst displayed in **Figure 7**. It should be noted that the nature of the coke deposited on the deactivated catalyst at 123 min reaction is amorphous and fibrillary. Ochoa et al. [74] demonstrated that coke content evolves with time on stream. In this evolution, both types of cokes growth throughout time on stream with different rates, with the one for filamentous coke being higher. However, it can be concluded that coke deposition is more influenced by the growth of encapsulating coke, as has also been established in the literature [75, 76].

3.2 Pyrolysis and in-line reforming of different plastics

Approximately 40% of the post-consumption wastes come from agriculture, construction, automotive and distribution sectors, and the remaining 60% from municipal solid wastes (MSW). Although the plastics only account for 7 wt% in the MSW, its contribution is 20 vol% of the total MSW volume. Therefore, the

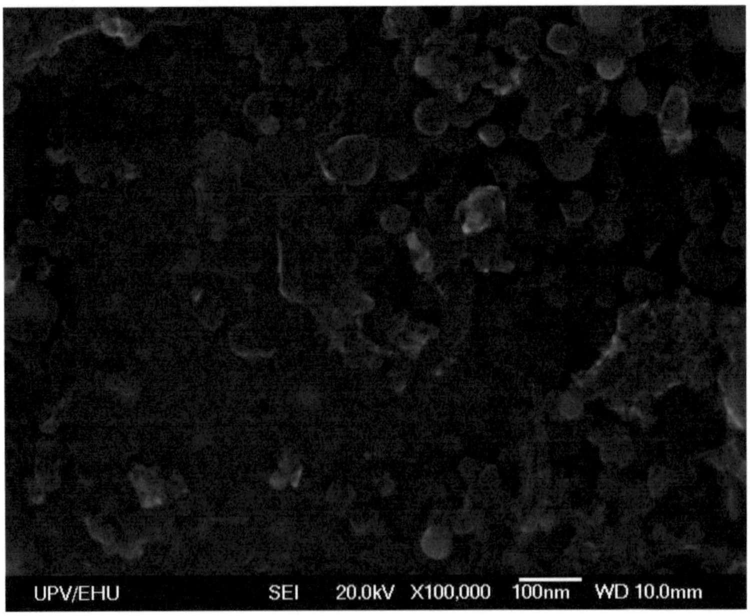

Figure 7.
SEM image of the deactivated catalyst.

thermochemical routes are of especial relevance for plastic valorization for obtaining fuels or for the production of chemical products, such as hydrogen. Moreover, the large scale viability of these processes will presumably require the joint valorization of different plastics from MSW. Accordingly, this section deals with the valorization of plastics of different nature by means of pyrolysis and in-line reforming strategy in order to evaluate its feasibility. Based on their production scale, the following plastics were selected: high density polyethylene (HDPE), polypropylene (PP), polyethylene terephthalate (PET) and polystyrene (PS). The operating conditions selected to carry out this study have been as follows: (i) temperature of the pyrolysis step, 500°C; (ii) temperature of the reforming step, 700°C; (iii) space time, 16.7 $g_{catalyst}$ min g_{HDPE}^{-1}; (iv) S/P ratio of 4.

The hydrogen produced in the pyrolysis and in-line catalytic reforming of different plastics is shown in **Figure 8a**. As observed, the highest hydrogen production is obtained when HDPE and PP are used as feed, 37.3 and 34.8 wt%, respectively, followed by PS (29.1 wt%), which has a lower hydrogen content in its composition. However, the oxygenated nature of the PET and the residue formed in its pyrolysis step leads to a lower hydrogen production in the reforming step (18.2 wt%). Wu and Williams [26] also observed lower conversion of PS pyrolysis volatiles compared to HDPE in the pyrolysis and in-line reforming of different plastics. A hydrogen production of 34 wt% was obtained by Czernik and French [31] in the pyrolysis and in-line reforming of PP using two fluidized bed reactors. Namioka et al. [29] performed the pyrolysis-reforming of PP and PS, with hydrogen production being 36 and 33 wt%, respectively. Recently, Saad and Williams [28] studied the pyrolysis and in-line dry reforming of different plastics, obtaining hydrogen productions of 15, 7.6 and 2.5 wt% for PE, PS and PET, respectively.

It should be highlighted that the Ni catalyst used for the pyrolysis-reforming process allows achieving a high conversion of pyrolysis products to gases (>98%), with a space time of 16.7 $g_{catalyst}$ min $g_{plastic}^{-1}$. This result is comparable to those obtained in the literature on Ru catalysts. Park et al. [30] obtained a similar conversion in the pyrolysis-reforming of PP, in a continuous regime system consisting of fixed bed reactors at 400 and 630°C, respectively, and using a space time of 9.8 $g_{catalyst}$ min $g_{plastic}^{-1}$. Based on this result, these authors note that the advantage of using Ru catalysts instead of Ni is the reduction of the reforming temperature by 70°C. Namioka et al. [29] extended the study using Ru catalysts to the pyrolysis-reforming of PS, verifying that under the indicated conditions the differences

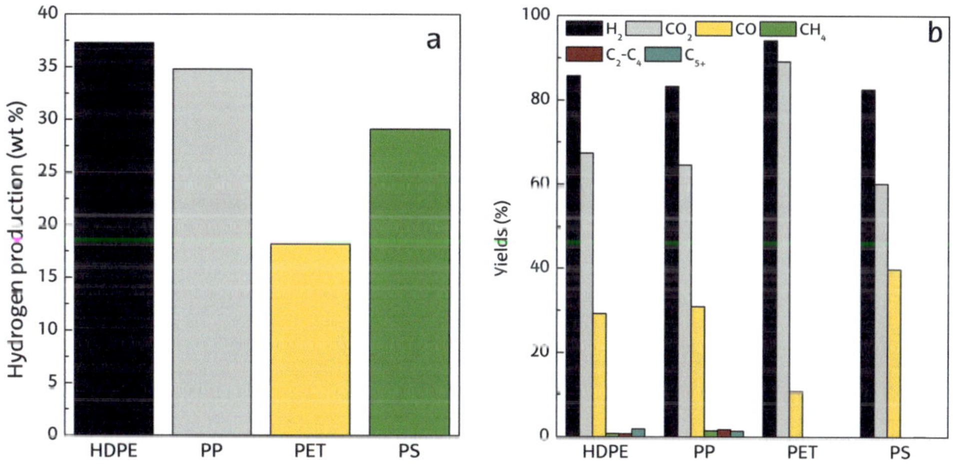

Figure 8.
Hydrogen production (a) and product yields (b) in the pyrolysis and in-line reforming of different plastics.

between the results for PP and PS are small. However, the most used catalysts in reforming processes are those based on Ni and Co, due mainly to its low cost in relation to noble metals.

Figure 8b shows the product yields obtained in the reforming of plastic pyrolysis volatiles. It should be noted that the hydrogen yield is based on the maximum allowable by stoichiometry, but those of the other compounds are given by C mole unit fed into the reforming step. As observed, there is a considerable effect on H_2, CO_2 and CO yields. On the one hand, the product yields are similar for HDPE and PP, as a consequence of the similar composition of the pyrolysis volatiles to be reformed, with hydrogen yields being 85.7 for HDPE and 83.3% for PP, respectively, and CO_2 yields of 67.4 and 64.6% and CO yields of 29.3 and 30.9%, respectively.

However, the distribution of the reforming products is different in the case of PET due to the differences in the pyrolysis volatile stream, with CO_2 and CO yields being of 29.3 and 9.9%, respectively. Thus, the product distribution in the reforming step is as follows: hydrogen yield 94.1%, CO_2 yield 89.2% and CO yield 10.8%. The high hydrogen yield and low CO yield obtained allow discarding the hypothesis that the origin of the CO_2 is the thermal decomposition of the oxygenate compounds. It is in fact produced by the WGS reaction, which is enhanced by the higher value of the effective space time (by carbon atom unit) for this plastic. In addition, it should be noted that yields of CH_4, C_2—C_4 hydrocarbons and non-converted C_{5+} are lower than in the case of HDPE and PP, which is a consequence of the higher effective space time used with this plastic.

On the other hand, the hydrogen yield obtained when PS is fed is similar to that obtained in the cases of HDPE and PP, whereas the CO_2 yield is slightly lower (60.2%) and that of CO slightly higher (39.8%). This result may also be associated with the lower extent of the WGS reaction due to the lower effective space time used in the case of this plastic.

It should also be pointed out that the results obtained for the PP and HDPE are similar to those obtained in the literature using Ru catalysts [29, 30], although they are obtained at lower temperatures (630 vs. 700°C) and using lower space time values (9.8 vs. 16.7 $g_{catalyst}$ min $g_{plastic}^{-1}$).

4. Conclusions

The hydrogen production from plastic has been studied in a two-step process. The first is the polymer fast pyrolysis and the second the catalytic steam reforming of pyrolysis volatiles. The process has been developed in a bench scale unit operating in continuous regime. The conical spouted bed reactor has been selected for the pyrolysis step because its features avoid operational problems and allow operating at low pyrolysis temperatures (500°C), minimizing the energy requirement. It has been observed that the yields and the composition of pyrolysis volatiles obtained at 500°C feeding steam are similar to those corresponding to the pyrolysis under N_2 atmosphere.

The temperature, space time and steam/plastic ratio in the reforming step have a great influence on the product yields and distributions obtained. An increase in the three variables favors the reforming of the pyrolysis volatiles, obtaining a hydrogen yield of 92.5% of that allowable by stoichiometry for a space time of 16.7 $g_{catalyst}$ min g_{HDPE}^{-1} at 700°C when a steam/plastic ratio of 5 is used. An increase in temperature in the 600–700°C range increases the conversion from 94.5 to 98.1% and the hydrogen yield from 78.1 to 85.7%. However, the WGS reaction is hindered, which increases the CO yield. An increase in space time causes a considerable increase in the conversion and yields of H_2, CO_2 and CO, and a decrease in the yields of CH_4, C_2—C_4 and C_{5+} hydrocarbons.

Moreover, an increase in the steam/plastic ratio increases the hydrogen yield, by displacing the reforming and the WGS reactions. The conversion obtained with a steam/plastic ratio of 5 at 700°C and with a space time of 16.7 $g_{catalyst}$ min g_{HDPE}^{-1} is almost full (99.7%), with a hydrogen yield of 92.5%.

The use of a fluidized bed reactor in the reforming step allows the operation without problems. However, reaction rate decreases due to the deposition of coke on the catalyst. As a result of this deposition, the specific surface and pore volume of the catalyst decrease and the yields of non-reformed hydrocarbons (CH_4, light hydrocarbons (C_2—C_4) and C_{5+} fraction) increase with time on stream.

The individual valorization of the different plastics leads to conversion values above 98% in all cases. The reforming behavior is similar for HDPE and PP. PET leads to higher yields of H_2 (94.1%) and CO_2 (89.2%). In the case of PS, H_2 yield is similar to those obtained when HDPE and PP (82.5%) are fed, whereas the CO_2 yield is slightly lower (60.2%).

Acknowledgements

This work was carried out with financial support from the Ministry of Economy and Competitiveness of the Spanish Government (CTQ2016-75535-R and CTQ-2015-69436-R), the European Regional Development Funds (ERDF), the Basque Government (IT748-13) and the University of the Basque Country (UFI 11/39).

Conflict of interest

The authors of this chapter declare no conflict of interest.

Author details

Itsaso Barbarias, Aitor Arregi, Maite Artetxe, Laura Santamaria, Gartzen Lopez*, María Cortazar, Maider Amutio, Javier Bilbao and Martin Olazar
Department of Chemical Engineering, University of the Basque Country (UPV/EHU), Bilbao, Spain

*Address all correspondence to: gartzen.lopez@ehu.es

References

[1] Arregi A, Amutio M, Lopez G, Bilbao J, Olazar M. Evaluation of thermochemical routes for hydrogen production from biomass: A review. Energy Conversion and Management. 2018;**165**:696-719. DOI: 10.1016/j.enconman.2018.03.089

[2] International Energy Agency (IEA). Technology Roadmap: Hydrogen and Fuel Cells; Paris. 2015

[3] Turner J, Sverdrup G, Mann MK, Maness P, Kroposki B, Ghirardi M, et al. Renewable hydrogen production. International Journal of Energy Research. 2008;**32**:379-407. DOI: 10.1002/er.1372

[4] Balat H, Kirtay E. Hydrogen from biomass—Present scenario and future prospects. International Journal of Hydrogen Energy. 2010;**35**:7416-7426. DOI: 10.1016/j.ijhydene.2010.04.137

[5] Trane R, Dahl S, Skjoth-Rasmussen MS, Jensen AD. Catalytic steam reforming of bio-oil. International Journal of Hydrogen Energy. 2012;**37**:6447-6472. DOI: 10.1016/j.ijhydene.2012.01.023

[6] Lopez G, Artetxe M, Amutio M, Alvarez J, Bilbao J, Olazar M. Recent advances in the gasification of waste plastics. A critical overview. Renewable and Sustainable Energy Reviews. 2018;**82**:576-596. DOI: 10.1016/j.rser.2017.09.032

[7] Xiao R, Jin B, Zhou H, Zhong Z, Zhang M. Air gasification of polypropylene plastic waste in fluidized bed gasifier. Energy Conversion and Management. 2007;**48**:778-786. DOI: 10.1016/j.enconman.2006.09.004

[8] Arena U, Di Gregorio F, Amorese C, Mastellone ML. A techno-economic comparison of fluidized bed gasification of two mixed plastic wastes. Waste Management. 2011;**31**:1494-1504

[9] Arena U, Di Gregorio F. Energy generation by air gasification of two industrial plastic wastes in a pilot scale fluidized bed reactor. Energy. 2014;**68**:735-743. DOI: 10.1016/j.energy.2014.01.084

[10] Cho M, Mun T, Choi Y, Kim J. Two-stage air gasification of mixed plastic waste: Olivine as the bed material and effects of various additives and a nickel-plated distributor on the tar removal. Energy. 2014;**70**:128-134. DOI: 10.1016/j.energy.2014.03.097

[11] Toledo JM, Aznar MP, Sancho JA. Catalytic air gasification of plastic waste (polypropylene) in a fluidized bed. Part II: Effects of some operating variables on the quality of the raw gas produced using olivine as the in-bed material. Industrial and Engineering Chemistry Research. 2011;**50**: 11815-11821. DOI: 10.1021/ie071023q

[12] Jeong Y, Choi Y, Kim J. Three-stage air gasification of waste polyethylene: In-situ regeneration of active carbon used as a tar removal additive. Energy. 2019;**166**:335-342. DOI: 10.1016/j.energy.2018.10.086

[13] Wilk V, Hofbauer H. Conversion of mixed plastic wastes in a dual fluidized bed steam gasifier. Fuel. 2013;**107**: 787-799. DOI: 10.1016/j.fuel.2013.01.068

[14] Kern SJ, Pfeifer C, Hofbauer H. Cogasification of polyethylene and lignite in a dual fluidized bed gasifier. Industrial and Engineering Chemistry Research. 2013;**52**:4360-4371

[15] Wilk V, Hofbauer H. Co-gasification of plastics and biomass in a dual fluidized-bed steam gasifier: Possible interactions of fuels. Energy and Fuels. 2013;**27**:3261-3273. DOI: 10.1021/ef400349k

[16] Friengfung P, Jamkrajang E, Sunphorka S, Kuchonthara P, Mekasut L.

NiO/dolomite catalyzed steam/O_2 gasification of different plastics and their mixtures. Industrial and Engineering Chemistry Research. 2014;**53**:1909-1915. DOI: 10.1021/ie401893s

[17] Baloch HA, Yang T, Li R, Nizamuddin S, Kai X, Bhutto AW. Parametric study of co-gasification of ternary blends of rice straw, polyethylene and polyvinylchloride. Clean Technologies and Environmental Policy. 2016;**18**:1031-1042. DOI: 10.1007/s10098-016-1092-4

[18] Ahmed II, Gupta AK. Hydrogen production from polystyrene pyrolysis and gasification: Characteristics and kinetics. International Journal of Hydrogen Energy. 2009;**34**:6253-6264. DOI: 10.1016/j.ijhydene.2009.05.046

[19] Guo X, Zhang W, Wang L, Hao J. Comparative study of nitrogen migration among the products from catalytic pyrolysis and gasification of waste rigid polyurethane foam. Journal of Analytical and Applied Pyrolysis. 2016;**120**:144-153. DOI: 10.1016/j.jaap.2016.04.018

[20] Erkiaga A, Lopez G, Amutio M, Bilbao J, Olazar M. Syngas from steam gasification of polyethylene in a conical spouted bed reactor. Fuel. 2013;**109**:461-469. DOI: 10.1016/j.fuel.2013.03.022

[21] Lopez G, Erkiaga A, Artetxe M, Amutio M, Bilbao J, Olazar M. Hydrogen production by high density polyethylene steam gasification and in-line volatile reforming. Industrial and Engineering Chemistry Research. 2015;**54**:9536-9544. DOI: 10.1021/acs.iecr.5b02413

[22] Lopez G, Erkiaga A, Amutio M, Bilbao J, Olazar M. Effect of polyethylene co-feeding in the steam gasification of biomass in a conical spouted bed reactor. Fuel. 2015;**153**:393-401. DOI: 10.1016/j.fuel.2015.03.006

[23] Rutberg PG, Kuznetsov VA, Serba EO, Popov SD, Surov AV, Nakonechny GV, et al. Novel three-phase steam-air plasma torch for gasification of high-caloric waste. Applied Energy. 2013;**108**:505-514. DOI: 10.1016/j.apenergy.2013.03.052

[24] Hlina M, Hrabovsky M, Kavka T, Konrad M. Production of high quality syngas from argon/water plasma gasification of biomass and waste. Waste Management. 2014;**34**:63-66. DOI: 10.1016/j.wasman.2013.09.018

[25] Wu C, Williams PT. A novel Ni-Mg-Al-CaO catalyst with the dual functions of catalysis and CO_2 sorption for H_2 production from the pyrolysis-gasification of polypropylene. Fuel. 2010;**89**:1435-1441. DOI: 10.1016/j.fuel.2009.10.020

[26] Wu C, Williams PT. Investigation of coke formation on Ni-Mg-Al catalyst for hydrogen production from the catalytic steam pyrolysis-gasification of polypropylene. Applied Catalysis B: Environmental. 2010;**96**:198-207. DOI: 10.1016/j.apcatb.2010.02.022

[27] Wu C, Williams PT. Investigation of Ni-Al, Ni-Mg-Al and Ni-Cu-Al catalyst for hydrogen production from pyrolysis-gasification of polypropylene. Applied Catalysis B: Environmental. 2009;**90**:147-156. DOI: 10.1016/j.apcatb.2009.03.004

[28] Saad JM, Williams PT. Pyrolysis-catalytic-dry reforming of waste plastics and mixed waste plastics for syngas production. Energy and Fuels. 2016;**30**:3198-3204. DOI: 10.1021/acs.energyfuels.5b02508

[29] Namioka T, Saito A, Inoue Y, Park Y, Min TJ, Roh SA, et al. Hydrogen-rich gas production from waste plastics by pyrolysis and low-temperature steam reforming over a ruthenium catalyst. Applied Energy. 2011;**88**:2019-2026. DOI: 10.1016/j.apenergy.2010.12.053

[30] Park Y, Namioka T, Sakamoto S, Min TJ, Roh SA, Yoshikawa K. Optimum operating conditions for a two-stage gasification process fueled by polypropylene by means of continuous reactor over ruthenium catalyst. Fuel Processing Technology. 2010;**91**:951-957. DOI: 10.1016/j.fuproc.2009.10.014

[31] Czernik S, French RJ. Production of hydrogen from plastics by pyrolysis and catalytic steam reform. Energy and Fuels. 2006;**20**:754-758. DOI: 10.1016/j.fuproc.2009.10.014

[32] Erkiaga A, Lopez G, Barbarias I, Artetxe M, Amutio M, Bilbao J, et al. HDPE pyrolysis-steam reforming in a tandem spouted bed-fixed bed reactor for H_2 production. Journal of Analytical and Applied Pyrolysis. 2015;**116**:34-41. DOI: 10.1016/j.jaap.2015.10.010

[33] Barbarias I, Lopez G, Alvarez J, Artetxe M, Arregi A, Bilbao J, et al. A sequential process for hydrogen production based on continuous HDPE fast pyrolysis and in-line steam reforming. Chemical Engineering Journal. 2016;**296**:191-198. DOI: 10.1016/j.cej.2016.03.091

[34] Barbarias I, Lopez G, Artetxe M, Arregi A, Santamaria L, Bilbao J, et al. Pyrolysis and in-line catalytic steam reforming of polystyrene through a two-step reaction system. Journal of Analytical and Applied Pyrolysis. 2016;**122**:502-510. DOI: 10.1016/j.jaap.2016.10.006

[35] Barbarias I, Lopez G, Artetxe M, Arregi A, Bilbao J, Olazar M. Valorisation of different waste plastics by pyrolysis and in-line catalytic steam reforming for hydrogen production. Energy Conversion and Management. 2018;**156**:575-584. DOI: 10.1016/j.enconman.2017.11.048

[36] Arregi A, Amutio M, Lopez G, Artetxe M, Alvarez J, Bilbao J, et al. Hydrogen-rich gas production by continuous pyrolysis and in-line catalytic reforming of pine wood waste and HDPE mixtures. Energy Conversion and Management. 2017;**136**:192-201. DOI: 10.1016/j.enconman.2017.01.008

[37] Al-Salem SM, Lettieri P, Baeyens J. Recycling and recovery routes of plastic solid waste (PSW): A review. Waste Management. 2009;**29**:2625-2643. DOI: 10.1016/j.wasman.2009.06.004

[38] Lopez G, Artetxe M, Amutio M, Bilbao J, Olazar M. Thermochemical routes for the valorization of waste polyolefinic plastics to produce fuels and chemicals. A review. Renewable and Sustainable Energy Reviews. 2017;**73**:346-368. DOI: 10.1016/j.rser.2017.01.142

[39] Wong SL, Ngadi N, Abdullah TAT, Inuwa IM. Current state and future prospects of plastic waste as source of fuel: A review. Renewable and Sustainable Energy Reviews. 2015;**50**:1167-1180. DOI: 10.1016/j.rser.2015.04.063

[40] Anuar Sharuddin SD, Abnisa F, Wan Daud WMA, Aroua MK. A review on pyrolysis of plastic wastes. Energy Conversion and Management. 2016;**115**:308-326. DOI: 10.1016/j.enconman.2016.02.037

[41] Zhang X, Lei H, Chen S, Wu J. Catalytic co-pyrolysis of lignocellulosic biomass with polymers: A critical review. Green Chemistry. 2016;**18**:4145-4169. DOI: 10.1039/C6GC00911E

[42] Jung SH, Cho MH, Kang BS, Kim JS. Pyrolysis of a fraction of waste polypropylene and polyethylene for the recovery of BTX aromatics using a fluidized bed reactor. Fuel Processing Technology. 2010;**91**:277-284. DOI: 10.1016/j.fuproc.2009.10.009

[43] Kaminsky W, Predel M, Sadiki A. Feedstock recycling of polymers by pyrolysis in a fluidised bed. Polymer Degradation and Stability.

2004;**85**:1045-1050. DOI: 10.1016/j.polymdegradstab.2003.05.002

[44] Kaminsky W, Schlesselmann B, Simon C. Olefins from polyolefins and mixed plastics by pyrolysis. Journal of Analytical and Applied Pyrolysis. 1995;**32**:19-27. DOI: 10.1016/0165-2370(94)00830-T

[45] Kaminsky W, Schlesselmann B, Simon CM. Thermal degradation of mixed plastic waste to aromatics and gas. Polymer Degradation and Stability. 1996;**53**:189-197. DOI: 10.1016/0141-3910(96)00087-0

[46] Mastellone ML, Arena U. Bed defluidisation during the fluidised bed pyrolysis of plastic waste mixtures. Polymer Degradation and Stability. 2004;**85**:1051-1058. DOI: 10.1016/j.polymdegradstab.2003.04.002

[47] Artetxe M, Lopez G, Amutio M, Elordi G, Olazar M, Bilbao J. Operating conditions for the pyrolysis of poly-(ethylene terephthalate) in a conical spouted-bed reactor. Industrial and Engineering Chemistry Research. 2010;**49**:2064-2069. DOI: 10.1021/ie900557c

[48] Arabiourrutia M, Elordi G, Lopez G, Borsella E, Bilbao J, Olazar M. Characterization of the waxes obtained by the pyrolysis of polyolefin plastics in a conical spouted bed reactor. Journal of Analytical and Applied Pyrolysis. 2012;**94**:230-237. DOI: 10.1016/j.jaap.2011.12.012

[49] Elordi G, Olazar M, Lopez G, Artetxe M, Bilbao J. Product yields and compositions in the continuous pyrolysis of high-density polyethylene in a conical spouted bed reactor. Industrial and Engineering Chemistry Research. 2011;**50**:6650-6659. DOI: 10.1021/ie200186m

[50] Elordi G, Olazar M, Lopez G, Artetxe M, Bilbao J. Continuous polyolefin cracking on an HZSM-5 zeolite catalyst in a conical spouted bed reactor. Industrial and Engineering Chemistry Research. 2011;**50**:6061-6070. DOI: 10.1021/ie2002999

[51] Lopez G, Alvarez J, Amutio M, Mkhize NM, Danon B, van der Gryp P, et al. Waste truck-tyre processing by flash pyrolysis in a conical spouted bed reactor. Energy Conversion and Management. 2017;**142**:523-532. DOI: 10.1016/j.enconman.2017.03.051

[52] Artetxe M, Lopez G, Amutio M, Barbarias I, Arregi A, Aguado R, et al. Styrene recovery from polystyrene by flash pyrolysis in a conical spouted bed reactor. Waste Management. 2015;**46**:126-133. DOI: 10.1016/j.wasman.2015.05.034

[53] Fernandez-Akarregi AR, Makibar J, Lopez G, Amutio M, Olazar M. Design and operation of a conical spouted bed reactor pilot plant (25 kg/h) for biomass fast pyrolysis. Fuel Processing Technology. 2013;**112**:48-56. DOI: 10.1016/j.fuproc.2013.02.022

[54] Makibar J, Fernandez-Akarregi AR, Amutio M, Lopez G, Olazar M. Performance of a conical spouted bed pilot plant for bio-oil production by poplar flash pyrolysis. Fuel Processing Technology. 2015;**137**:283-289. DOI: 10.1016/j.fuproc.2015.03.011

[55] Al-Salem SM, Antelava A, Constantinou A, Manos G, Dutta A. A review on thermal and catalytic pyrolysis of plastic solid waste (PSW). Journal of Environmental Management. 2017;**197**:177-198. DOI: 10.1016/j.jenvman.2017.03.084

[56] Kunwar B, Cheng HN, Chandrashekaran SR, Sharma BK. Plastics to fuel: A review. Renewable and Sustainable Energy Reviews. 2016;**54**:421-428. DOI: 10.1016/j.rser.2015.10.015

[57] Wu C, Williams PT. Pyrolysis-gasification of post-consumer municipal solid plastic waste for hydrogen production. International Journal of Hydrogen Energy. 2010;**35**:949-957. DOI: 10.1016/j.ijhydene.2009.11.045

[58] Tsuji T, Hatayama A. Gasification of waste plastics by steam reforming in a fluidized bed. Journal of Material Cycles and Waste Management. 2009;**11**:144-147. DOI: 10.1007/s10163-008-0227-z

[59] Luo S, Zhou Y, Yi C. Syngas production by catalytic steam gasification of municipal solid waste in fixed-bed reactor. Energy. 2012;**44**:391-395. DOI: 10.1016/j.energy.2012.06.016

[60] Wu C, Williams PT. Effects of gasification temperature and catalyst ratio on hydrogen production from catalytic steam pyrolysis-gasification of polypropylene. Energy and Fuels. 2008;**22**:4125-4132. DOI: 10.1021/ef800574w

[61] Wang J, Cheng G, You Y, Xiao B, Liu S, He P, et al. Hydrogen-rich gas production by steam gasification of municipal solid waste (MSW) using NiO supported on modified dolomite. International Journal of Hydrogen Energy. 2012;**37**:6503-6510. DOI: 10.1016/j.ijhydene.2012.01.070

[62] Li J, Liao S, Dan W, Jia K, Zhou X. Experimental study on catalytic steam gasification of municipal solid waste for bioenergy production in a combined fixed bed reactor. Biomass and Bioenergy. 2012;**46**:174-180. DOI: 10.1016/j.biombioe.2012.08.026

[63] Acomb JC, Wu C, Williams PT. Control of steam input to the pyrolysis-gasification of waste plastics for improved production of hydrogen or carbon nanotubes. Applied Catalysis B: Environmental. 2014;**147**:571-584. DOI: 10.1016/j.apcatb.2013.09.018

[64] Wu C, Williams PT. Hydrogen production from the pyrolysis/gasification of polypropylene: Influence of steam flow rate, carrier gas flow rate and gasification temperature. Energy and Fuels. 2009;**23**:5055-5061. DOI: 10.1021/ef900278w

[65] Wu C, Williams PT. Pyrolysis-gasification of plastics, mixed plastics and real-world plastic waste with and without Ni-Mg-Al catalyst. Fuel. 2010;**89**:3022-3032. DOI: 10.1016/j.fuel.2010.05.032

[66] Donatelli A, Iovane P, Molino A. High energy syngas production by waste tyres steam gasification in a rotary kiln pilot plant. Experimental and numerical investigations. Fuel. 2010;**89**:2721-2728. DOI: 10.1016/j.fuel.2010.03.040

[67] Barbarias I, Artetxe M, Lopez G, Arregi A, Bilbao J, Olazar M. Influence of the conditions for reforming HDPE pyrolysis volatiles on the catalyst deactivation by coke. Fuel Processing Technology. 2018;**171**:100-109. DOI: 10.1016/j.fuproc.2017.11.003

[68] Remiro A, Valle B, Aguayo AT, Bilbao J, Gayubo AG. Operating conditions for attenuating Ni/La_2O_3-Al_2O_3 catalyst deactivation in the steam reforming of bio-oil aqueous fraction. Fuel Processing Technology. 2013;**115**:222-232. DOI: 10.1016/j.fuproc.2013.06.003

[69] Vicente J, Montero C, Ereña J, Azkoiti MJ, Bilbao J, Gayubo AG. Coke deactivation of Ni and Co catalysts in ethanol steam reforming at mild temperatures in a fluidized bed reactor. International Journal of Hydrogen Energy. 2014;**39**:12586-12596. DOI: 10.1016/j.ijhydene.2014.06.093

[70] Rostrup-Nielsen JR, Sehested J. Whisker carbon revisited. Studies

in Surface Science and Catalysis. 2001;**139**:1-12. DOI: 10.1016/S0167-2991(01)80174-9

[71] Joensen F, Rostrup-Nielsen JR. Conversion of hydrocarbons and alcohols for fuel cells. Journal of Power Sources. 2002;**105**:195-201. DOI: 10.1016/S0378-7753(01)00939-9

[72] Angeli SD, Pilitsis FG, Lemonidou AA. Methane steam reforming at low temperature: Effect of light alkanes' presence on coke formation. Catalysis Today. 2015:119-128. DOI: 10.1016/j.cattod.2014.05.043

[73] Barbarias I, Lopez G, Amutio M, Artetxe M, Alvarez J, Arregi A, et al. Steam reforming of plastic pyrolysis model hydrocarbons and catalyst deactivation. Applied Catalysis A: General. 2016;**527**:152-160. DOI: 10.1016/j.apcata.2016.09.003

[74] Ochoa A, Barbarias I, Artetxe M, Gayubo AG, Olazar M, Bilbao J, et al. Deactivation dynamics of a Ni supported catalyst during the steam reforming of volatiles from waste polyethylene pyrolysis. Applied Catalysis B: Environmental. 2017;**209**:554-565. DOI: 10.1016/j.apcatb.2017.02.015

[75] Montero C, Ochoa A, Castaño P, Bilbao J, Gayubo AG. Monitoring NiO and coke evolution during the deactivation of a Ni/La_2O_3-αAl_2O_3 catalyst in ethanol steam reforming in a fluidized bed. Journal of Catalysis. 2015;**331**:181-192. DOI: 10.1016/j.jcat.2015.08.005

[76] Trimm DL. Coke formation and minimisation during steam reforming reactions. Catalysis Today. 1997;**37**:233. DOI: 10.1016/S0920-5861(97)00014-X

Chapter 6

Synthesis and Characterization of Forsterite (Mg_2SiO_4) Nanomaterials of Dunite from Sumatera

Ratnawulan Ratnawulan and Ahmad Fauzi

Abstract

The topics to be discussed are about the synthesis and characterization of forsterite nanoparticles (Mg_2So_4) from dunite rocks originating from West Sumatra. This region is a meeting of two Indian and Australian plates that give rise to mineral dunite with unique characteristics. Forsterite synthesis was carried out using calcinations temperature variations namely 700, 800, 900, 1000 and 1100°C. Synthesis of forsterite nanoparticles used the High Energy Milling Ellipse 3D Motion (HEM-E3D) method with variations in milling time, i.e., 5, 10, 20 h. Characterization was carried out using X-Ray Fluorescence (XRF) and X-Ray Diffraction (XRD) and Scanning Electron Microscopy (SEM). The synthesis results explain the forsterite concentration in dunite rock, pure forsterite at optimal calcinations temperature, forsterite nanoparticles at optimal grinding time and forsterite nanoparticle structure.

Keywords: forsterite, nanoparticles, dunite, HEM-E3D, milling time

1. Introduction

Forsterite is a magnesium silicate crystal with the chemical formula Mg_2SiO_4 derived from the olivine mineral group [1]. Olivine is a group of minerals composed of iron (Fe) and magnesium (Mg). The olivine mineral is green, with luster, formed at high temperatures. This mineral is commonly found in basalt and ultramafic rocks. Rocks whose overall minerals consist of olivine minerals are known as dunite rocks [2, 3].

Forsterite is often used as an electronic component because it has a low coefficient of thermal expansion and electrical conductivity [4]. Besides that, forsterite has superior properties, namely chemical stability at high temperatures [5] and high electrical properties so that it can be used as a coating for the back, edges, and front of iron and steel [6–11].

Iron and steel are widely used in industry and property such as building materials, ships, electronic equipment, etc. But in its use, steel is easily corrosion. To overcome this corrosion problem, a lot of research continues. One of them is by soaking the steel with an anti-corrosion solution. But the results are not yet suitable for achieving commercial applications. The reason is that the particle size of the anti-corrosion powder used is relatively very large, giving rise to new problems, namely iron and steel which look uneven and homogeneous so that it disrupts appearance or esthetics.

Nanoparticle technology brings fresh air in an effort to increase the resistance of steel to corrosion. With a very small particle size, the problem can be solved. Nanoparticles are very fine so they are homogeneous. The use of nanoparticles is widely applied in various fields, including industrial fields such as paint, health, etc. Nanometer-sized materials have chemical and physical properties that are superior to large (bulk) material [12]. Nanoparticles have several advantages, namely having a larger touch surface area so that the bond between one particle and another is easy to form, and the mechanical, optical, and chemical properties of the material can experience significant differences with the properties of the sized material micrometer [13].

The advantage of forsterite made in nanometer size is that it is very strong, hard, and resilient at high temperatures, and is waterproof, corrosion resistant and has very active chemical properties that add durability when used for iron and steel coatings. Economical studies of nanoforsterite or nanomaterials are far more economical than conventional [14].

Forsterite nanoparticles can be made chemically and mechanically. Chemical methods are co-precipitation, solvothermal, sol-gel, solid state, and others. The mechanical method is by using a ball mill [15]. The advantage of using mechanical milling is a simple and effective method for growing solid crystals (the size of crystal grains becomes small) without going through chemical reactions that require a long time in the process of nanoparticle synthesis [15].

Forsterite nanoparticle synthesis has been carried out using synthetic samples with the milling method [16–18]. The crystallite and grain size obtained from the synthesis of nanoforsterite can be seen in **Table 1**.

Table 1 shows the grain sizes of different forsterites resulting from several different research treatments. Size the forsterite nano single phase is formed after 40 h of milling and heating the temperature is 1000°C with a holding time of 1 h. The grain size of the Forster produced is a range between 0.1 and 2 μm [16]. The synthesis of forsterite nanoparticles was then continued using talc and magnesium oxide by the same method [17]. The variation in milling time is increased to 60 h with a heating temperature of 1200°C. The grain size obtained is less than 500 nm. Another study also uses a milling process with a variation of 5, 10, 20, and 30 h milling with temperature variations of 850–1100°C. Forsterite nanopowders were prepared from MgO and SiO_2 mixtures by using a high energy ball milling method. The results found that the average particle size was reduced to 147.4 nm [18].

Methods	Ref.	Milling Time	Temperature (°C)	Crystalite Size (nm)	Grain Size
Milling	16	5 min- 40 h	1000	49	0.1 - 2 μm
Milling (continued using talc and magnesium oxide material)	17	1 min-60 h	1200	33	< 500 nm
Milling	18	5-30 h	850-1100	-	147.4 nm
High Energy Milling	Current Chapter	5-40 h	26-1100	18.78	345 nm

Table 1.
The crystallite and grain size obtained from the synthesis of nanoforsterite.

Characterization using XRD found that there was still Fe and Cr with a low content of 0.4256 and 0.5056%, respectively. Although many researchers have synthesized forsterite nanoparticles, the results obtained are still contaminated with other elements even at low levels. Contamination of samples with other elements can affect the nature of the final product.

This chapter discusses the synthesis and characterization of forsterite nanoparticles using High Energy Milling Ellipse 3D Motion (HEM-E3D). HEM-E3D is a unique technique that uses collision energy between crush balls and chamber walls which are rotated and moved in a certain way. The advantages of High Energy Milling is that in a relatively short time it can make nanoparticles (it takes several hours, depending on the type of tool), nanoparticles are produced in relatively large quantities. Besides that, the milling technique is one technique for growing solid crystals without going through the vaporation phase or chemical reaction treatment as is usually needed in the synthesis process in general.

2. West Sumatra and its geographical condition

The forsterite discussed in this chapter comes from dunite rocks originating from West Sumatra. West Sumatra is one of the regions in the Indonesian archipelago which has a fairly complex geological order. The West Sumatra region is traversed by the equator (zero degrees latitude), precisely located in the Bonjol subdistrict of Pasaman district, because of the influence of this location, West Sumatra has a trophic climate with high temperatures and humidity. The land surface height between one area and another area varies greatly. However, physically, West Sumatra is a region that is largely photographed by the mountains and the Bukit Barisan plateau which stretches from the North West to the Southeast, 63% and the area is a dense forest with elevations up to 3000 m above sea level.

The geographical condition of West Sumatra is quite unique which is partly located in the lowlands and partly in the highlands, marked there are many mountains, lakes, valleys/canyons, and rivers. This situation is due to its location at the confluence of three plates, namely the Eurasian plate to the north, the Australian Indian plate to the south and the Pacific plate to the east. The position of West Sumatra which is near the collision of two large plates namely the Australian Indian plate and the Eurasian Plate in addition to receiving negative consequences of natural disaster-prone areas also benefits, namely the emergence of sources of minerals containing economic minerals to the surface, one of which is dunite rock. **Figure 1** is dunite rock from West Sumatra.

Figure 1.
Dunite rocks from West Sumatra.

3. Dunite rock and its characteristics

Dunite is the main ingredient of the Earth's mantle and is rarely found in continental rock. Dunite is found when mantle rock plates from the subduction zone have been elevated to the continental crust. Dunite formation takes place in conditions that are dense or almost dense (at high temperatures) in a magma solution and before it cools to this temperature, the rock is ready to unite to form a binding anhedral olivine mass [19].

Dunite is a greenish-black rock and has a mineral composition almost entirely of monomineralic olivine (generally magnesia olivine) [20, 21]. Its mineral accessories include chromites, magnetite, limonite, and spinel. Olivine mineral is an iron-magnesium silicate with its chemical formula $(Mg, Fe)_2 (SiO_4)$, has an orthorhombic crystalline system, has no hemisphere, has a hardness of 6.5–7, specific gravity 3.27–4.37, has a gloss light, and the color of this mineral is yellowish green and grayish green. Olivine is one of the most common minerals on the surface of the earth and has also been identified on the surface of the Moon, Mars, and comets. Olivine minerals are generally as rock-forming minerals and also as accompanying minerals, in alkaline rocks such as gabbros, peridotite, etc. [21].

Dunite rocks naturally contain magnesium (MgO) and have a very alkaline pH, which is around 7.5–9.5. These rocks can be used as basic fertilizer (natural) and compound fertilizer. MgO is one of the most important elements for plants for photosynthesis, while the pH level of rocks can neutralize acid soils such as peatlands. Because the magnesium content of dunite rocks is quite high, the fertilizer with rock material will be very beneficial for agricultural and plantation activities.

Dunite contains 36–42% MgO and 36–39% SiO_2. Olivine is a commercial source of a combination of magnesia and silica used in metallurgy. The content will increase if it is affected by an increase in temperature. The phase diagram of dunite consists of two phases namely forsterite and fayalite. The characteristics of the forsterite phase are melted at 1890°C and fayalite melts at a temperature of 1205°C [22]. Ultramafic frozen rock is an igneous rock which chemically contains less than 45% SiO_2 from its composition. The mineral content is dominated by heavy minerals with elements such as Fe (iron) and Mg (magnesium) which are also called ultramafic minerals [23]. The mineral composition and structural characteristics of dunite mainly contain olivine minerals and always contain little brucite, spinel-chromite, magnetite, and pyroxene. For comparison, dunites from South Turkey have chemical compositions such as FeO: 7.05%, SiO_2: 38.74%, MgO: 37.16%, CaO: 9.24%, Al_2O_3: 1.65% [24].

4. The olivine mineral

Olivine is the name of a group of rock-forming minerals found in mafic and ultramafic igneous rocks such as basalt, gabbro, dunite, diabas, and peridotite. Olivine is usually green and has a chemical composition ranging from Mg_2SiO_4 and Fe_2SiO_4. Olivine has a high crystallization temperature compared to other minerals. Olivine is the first mineral that crystallizes from magma. Olivine crystals are formed during the slow cooling process of magma and then settle in the bottom of the magma kitchen because of its relatively high density. This olivine accumulation can cause dunite-like rock formation at the bottom of the magma kitchen.

The minerals in the olivine group crystallize in the orthorhombic system (Pbnm space group) with silicates, which means that olivine is nesosilicate. In an alternative view, the atomic structure can be described as hexagonal, near oxygen ions with half of the octahedral atoms bonded to magnesium or iron ions and

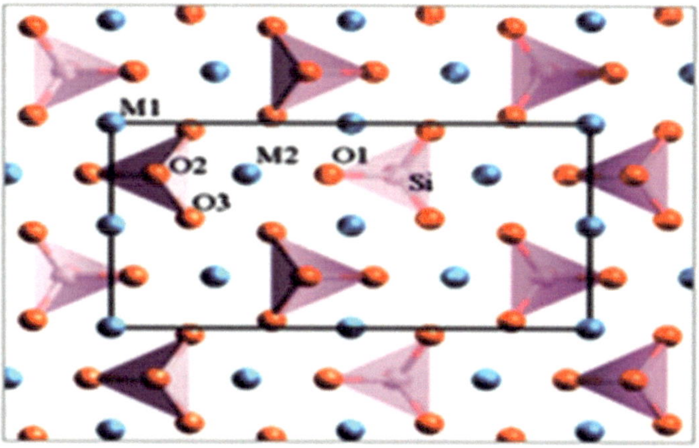

Figure 2.
Crystal structure of olivine [12].

one-eighth of tetrahedral occupied by silicon ions [12]. The shape of the olivine crystal structure can be seen in **Figure 2**.

5. Forsterite and its characteristics

Forsterite is a member of the olivine mineral prepared by Mg and Si. The general formula is Mg_2SiO_4. **Figure 3** shows the shape of the forsterite crystal structure.

Figure 3, Mg atoms are shown in purple, Mg2+ green, Si4+ white, and O2− pink. The radius of the Mg atom is 1.6 Å, Mg2+ is 0.75 Å, Si4+ is 0.4 Å, and O2 is 1.35 Å. A common characteristic of forsterite is the orthorhombic crystal system with the dimensions of cells a ≠ b ≠ c, where a = 4.79 Å; b = 10.19 Å; c = 5.85 Å with $\alpha = \beta = \gamma = 90°$ and space group Pbnm and density (g cm^{-3}) is 3.275 [26].

Forsterite has a low electrical conductivity making it ideal for electronic materials. In addition, forsterite has a high melting point which is equal to 1890°C

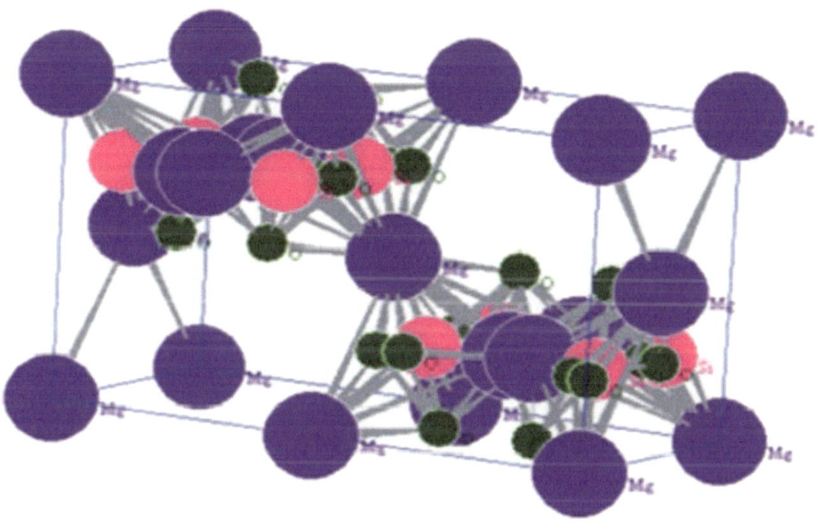

Figure 3.
Crystal structure of forsterite [25].

which indicates that forsterite is a refractory material and ceramic manufacturing application because it has good chemical stability and a low coefficient of thermal expansion [27].

6. Nanoparticles technology

The nanoparticle technology is a material technology that deals with the creation or synthesis of small particles in nanometers (one per billion meters). The purpose of this technology is to use it for a more efficient future life. Nanoparticle synthesis can be done in two ways, namely top down and bottom up. Top down is the making of nanostructures by minimizing large material by mechanical activation, while bottom up is a way of assembling atoms or molecules and combining them through chemical reactions to form nanostructures. An example of a bottom up technology is using sol gel techniques, chemical precipitation, and phase agglomeration while top down method is grinding with a milling tool such as HEM-E3D [15].

The process that occurs during milling can be explained below. Mechanical milling is a simple and effective method for growing solid crystals (the size of crystal grains becomes smaller) without going through the evaporation phase or chemical reaction, as is usually needed in other synthesis processes [25]. This smoothing machine is able to convert hard and easily broken samples into powder-shaped analytical samples [14].

HEM-E3D is one tool used for mechanical alloys that use hard balls made of carbon steel. HEM-E3D vibrates samples with engine shocks because milling is used for a collision of kinetic energy in the sample. The speed of the media causes a high amount of energy to form to collide with the sample. Various combinations of media milling are used in HEM. A ratio of balls to powder, or BPR (Ball Powder Ratio), usually also used to change milling parameters [28].

HEM-E3D is very good for reducing particle size. The nature of particle size reduction and subsequent growth is the same as oxide analysis. In the milling process, HEM-E3D works by destroying the powder mixture through the mechanism of colliding milled balls that move to follow the movement pattern of the three-dimensional elliptical container that allows the formation of nanometer scale powder particles due to the high frequency of collisions. The high frequency of collisions that occur between the mixture of powder and milled balls is caused by the container rotating at high speed, which reaches 500 rpm, and the shape of the spherical movement of the three-dimensional ellipse.

Milling with ball mill is used for pulverizing the powder into nanoparticles because there is a grinding of powder on the surface of the ball when it collides with another ball so that the impact size given by the ball is as big as the collision force of the surface area of the ball. There are two collision paths where the material is between two balls and pulverization where the material is exposed to the ball when it is close to the cylinder wall. The interaction between the balls in the chamber is shown in **Figure 4**.

Figure 4 shows the interaction between the balls in the chamber. The ball that has a smaller impact area, will give greater impact so that the destructive ability strengthens with the reduction of the touch area. Therefore, nanoparticle powder is easier to form by using smaller balls. Besides that, the collision frequency is the accelerating destruction factor.

The formation of forsterite nanoparticles using HEM-E3D will be more helpful because this tool has several advantages including machines that can be used can be used to do mixing, homogenization (uniformity), chemical mechanic (making chemical-mechanical reactions), mechanical milling, mechanical alloys, and do an

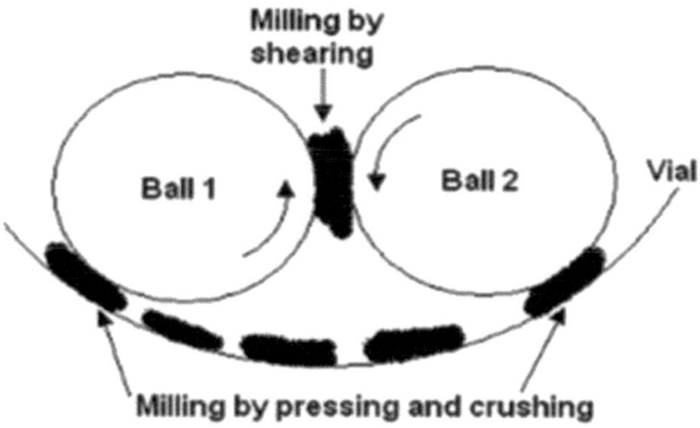

Figure 4.
Interaction between grinding balls in high energy milling [29].

emulsion. So nanoparticles will be produced by smoothing the material up to the nanometer scale, high destruction rate, easy conditioning of the milling system so that the mechanism of amorphization and nanoparticle formation process is faster and more effective, becoming a tool for making nanopowder at low prices and increasing efficiency and electronic systems integrated system includes a motor controller, and a timer.

Process parameters that must be considered in the grinding process include speed and time of grinding, comparison of ball weight to weight of powder, and empty space in the grinding container. In simple terms, increasing the rotation speed of the milling will increase the energy input to the powder. How fast the rotation of milling is affected by the design of the instrument. Speed also affects the increase in the temperature of the media milling. If the grinding speed is too high, the temperature of the components in the grinding process will increase. This increase in temperature can be advantageous for example when diffusion is needed to produce homogenizes and fusion of powder. The disadvantage is excessive friction or collision of the milling equipment that carries contamination.

The rotation speed used is adjusted so that the milling process runs optimally. If the rotation speed is set too fast, the balls will only rotate on the cylinder wall due to centrifugal force. Conversely, when the rotation speed is very low, the impact energy to destroy the material is not enough so that the process will last long.

The milling time needed also depends on the type of milling used, intensity of milling, ball-powder ratio, and milling temperature. Long milling times from the time needed will increase contamination and some unwanted phases will be formed. Therefore, milling powder did not require a long time for sample preparation.

In addition, the size of the ball greatly affects the efficiency of milling, where generally large ball sizes (with high density) are more useful because greater weight can transfer greater impact and kinetic energy to powder particles. While small balls produce more friction, making it easier to form an amorphous phase. The resulting grain size is also smaller when the ball used is small. In practice, a combination of various sizes is often used. The use of the same size of balls can cause the ball to rotate along the bullet path and not hit the surface of the powder randomly. The larger the size of the ball used, the greater the energy when pounding, but when the material is small, the collision will rarely occur because space, where the collision occurs, becomes smaller and in the end, the process will not be optimal.

For small scale or laboratory, generally, the ratio of balls to the weight of powder used is around 10:1, 10 g of balls and 1 g of powder, while for large scale

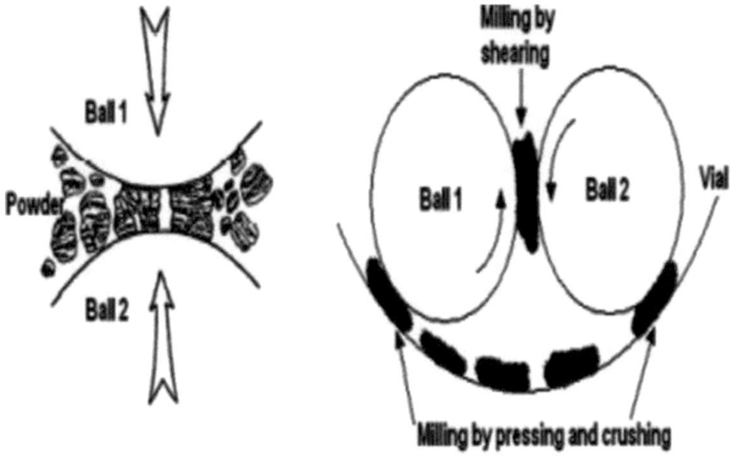

Figure 5.
Concept of powder destruction at ball mill [29].

or industrial scale, the ratio of ball weight to the weight of powder used can reach 100:1. The higher the ratio of the weight of the ball to the weight of the powder used, the shorter the frequency of impact.

The grinding procedure is a forsterite powder inserted into a metal chamber with several steel balls in it that move continuously. The size of the milled ball used in this process is a small milled ball with a diameter of 5 mm with a weight of 0.5 g. The ratio of the weight of the milled ball to the weight of the powder in the grinding container used is 10:1. For example, the weight of the milled ball is 150 g while the amount of powder is 15 g. In the metal chamber, the balls will collide with each other. As a result of this ball collision, the homogeneous powder that is inserted into this tool will be crushed between the balls. This causes the particles to break. And so on until it reaches the desired size [30]. **Figure 5** shows the concept of destruction of sample powder in the ball mill.

7. Synthesis of nanoparticles from forsterite

The synthesis of nanoparticles from forsterite discussed here comes from dunite rock samples taken from the West Sumatra region, precisely at coordinates: 00° 09′ 01.1″ LU and 99° 53′ 19.5″ BT. Samples were sieved with 2.1 mm sieves to obtain finer and homogeneous samples. The sample was then crushed for 3 min using bowl mill so that the samples were finer and homogeneous. Samples were tested by X-Ray Fluorescence (XRF) to see the composition of the compounds in the dunite. The results of XRF are known to dunite rocks found in West Sumatra, Indonesia has a composition of FeO: 11.12%, SiO_2: 31.18%, MgO: 49.96%, CaO: 7.31%, Al_2O_3: 0.43%. MgO content in this region is quite high when compared to Southern Turkey [24]. To get forsterite minerals from dunite samples, calcination was carried out with temperature variations of 700, 800, 900, 1000, and 1100°C. Then the samples were carried out X-Ray Diffraction (XRD) test to see the phases which appear for each calcination temperature. From the results of XRD, it is known that the forsterite phase at the optimum calcination temperature.

Samples with optimum calcination temperature are then taken to be further synthesized into forsterite nanoparticles. Samples were weighed as much as 2 g and milling balls as much as 20 g for each milling time. The type of milling ball used is small carbon steel with 40 pieces which weigh 0.2 per fruit. Medium sized milling balls of 4 which weigh 0.5 per piece. While large milling balls as much as two pieces weighing 3.55 g per fruit. The samples that have been prepared are then processed using HEM-E3D with variations in milling time of 5, 10, 20, 40, and 60 h. The length of the milling process in each cycle is carried out for 30 seconds, and then the process is stopped for 1 min to avoid an increase in temperature and damage to the milling device due to rising motor temperatures that are too high.

8. Calcination and the effect of calcination temperature on the raw material structure

The characterization of forsterite nanoparticles was carried out using X-ray Diffraction (XRD) and Scanning electron microscopes (SEM). XRD is used to determine the structure and size of crystals of forsterite for each variation of milling time. The crystal size of forsterite is calculated using the Scherrer formula, $D = 0.9\ \lambda/\beta \cos\theta$, where λ was the wavelength of X-ray radiation, β was the full width at half maximum (FWHM) of the peaks at the diffracting angle θ [31]. SEM is used to see the microstructure or morphology of forsterite and particle size. The results of the XRD test from dunite raw material for each calcination temperature variation can be seen in **Table 2**.

Table 1 shows the appearance of the forsterite phase starting at the calcination temperature of 700°C. But at this temperature, other phases are still found, namely fayalite, lizardite, magnesium iron silicate, olivine and enstatite. For the synthesis of forsterite into nanoparticles, the sample taken is the result of calcination at a temperature of 1100°C. The reason for taking this condition is the forsterite phase, even though the olivine phase is still found. Through the HEM-E3D method with variations in milling time, it is expected that the olivine phase can decompose into forsterite.

The XRD results for forsterite that have been milled for 5 h are shown in **Figure 6**.

Figure 6 is a plot of the 2θ diffraction angle on the intensity of the XRD pattern of forsterite after being processed for 5 h. The figure shows that the sample is

Calcinations' temperature (°C)	Phase	Structure				
		A(Å)	B	C	Space Grup	Crystal System
26	Lizardite	5.31	5.31	7.31	P31mE	Hexagon
	Magnetit	8.39	8.39	8.39	Fd-3m	Cubic
700	Fayalite	4.76	10.23	5.99	Pbnm	Orthorhon
	Forsterite	5.98	10.19	4.76	Pmnb	Orthorhon
800	Forsterite	4.76	10.21	5.99	Pbnm	Orthorhon
	Olivine	10.19	5.98	9.51	P1	Anorthi
	Lizardite	5.31	5.31	7.10	P31m	Hexagon
900	Forsterit	4.76	10.20	5.99	Pbnm	Orthorhon
	Magnesium Iron Silicate	4.76	10.22	5.99	Pbnm	Orthorhon
1000	Forsterit	4.76	10.21	5.98	Pbnm	Orthorhon
	Olivine	10.19	5.98	9.51	P1	Anorthi
1100	Enstatite	18.23	8.81	5.17	Pbca	Orthorhon
	Forsterite	4.75	10.19	5.98	Pbnm	Orthorhon
	Olivine	10.19	5.98	9.51	P1	Anorthi

Table 2.
Structure of dunite raw material for each temperature variation of calcination.

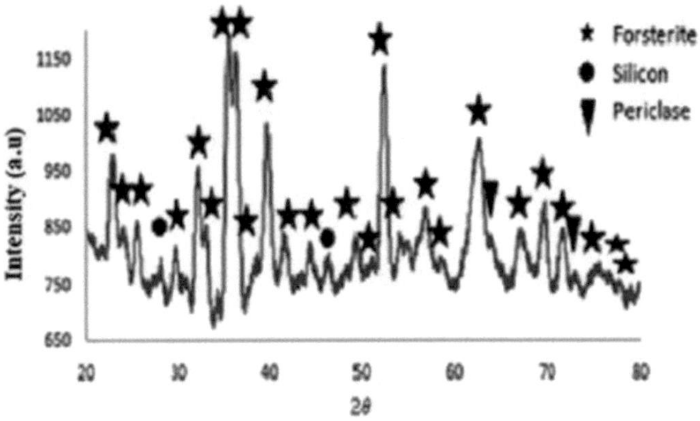

Figure 6.
XRD pattern from forsterite after being milled for 5 h.

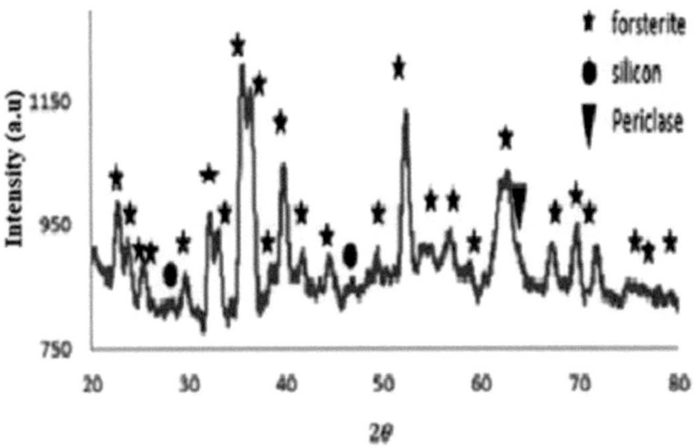

Figure 7.
XRD pattern from forsterite after being milled for 10 h.

dominated by the forsterite phase even though there are still other phases such as silicon which appear at scattering angles 2θ which are 26.63 and 47.2, and periclase with a scattering angle of 74.74 with a small degree.

Furthermore, for samples milled for 10 h, the XRD pattern is shown in **Figure 7**.

Figure 7 shows the sample is still contaminated with phases other than forsterite. Forsterite milled for 10 h there are several phases, namely forsterite, periclase, and silicon, but the phase concentration of the Periclase begins to decrease compared to samples which are milled for 5 h. The dominant phase is still forsterite.

Furthermore, for samples milled for 20 h, the XRD pattern is shown in **Figure 8**.

Figure 8 shows the sample is still contaminated with other phases other than forsterite. Forsterite milling for 20 h there are several phases, namely forsterite, periclase, and silicon. The dominant phase is still forsterite.

Furthermore, for samples milled for 40 h, the XRD pattern is shown in **Figure 9**.

Figure 9 shows no other phases such as periclase, and silicon that appear other than forsterite phase. This result is achieved after the sample has been milled for 40 h.

Based on XRD data analysis for each sample, it can be seen that variations in milling affect the phase of the sample. The phase that appears after being given the influence of milling variations for 5, 10, 20, and 40 h is shown in **Figure 10**.

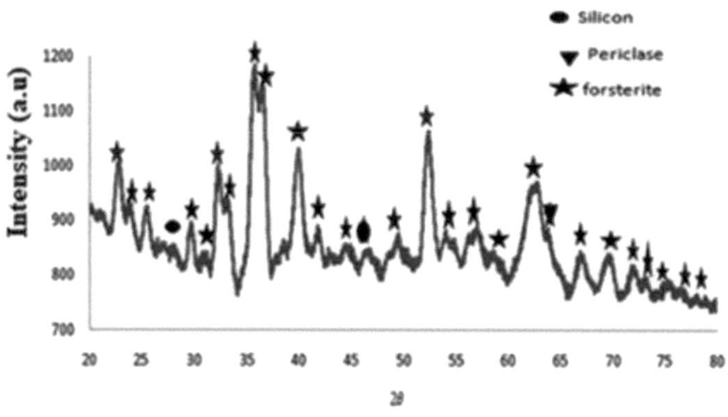

Figure 8.
XRD pattern of forsterite after being milled for 20 h.

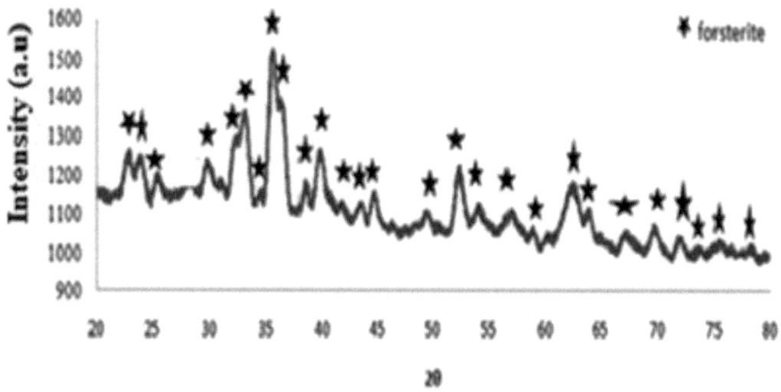

Figure 9.
XRD pattern from forsterite after being milled for 40 h.

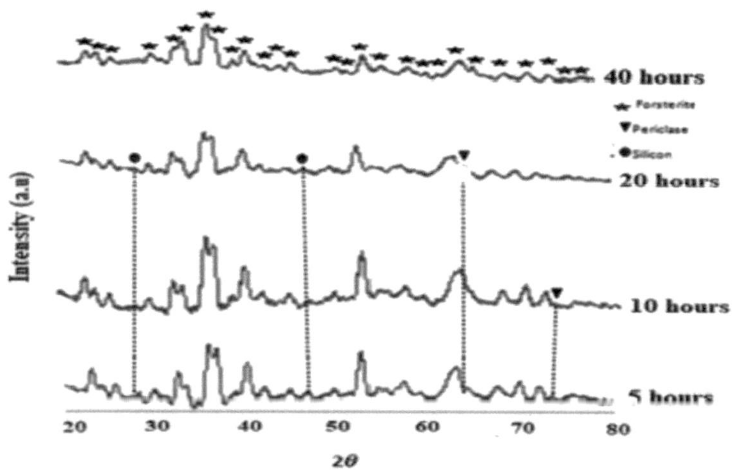

Figure 10.
Effect of milling time variation 5, 10, 20, and 40 h on the diffraction pattern of XRD results.

Figure 10 shows that at 5 h milling time, phases that appear forsterite, silicon, and periclase. The crystalline system at 5 h milling time is orthorhombic for the phosphoresce and cubic phases for the silicon and periclase phase. The crystallite size for the forsterite phase is 53.80 nm. While at the time of milling 10 h there was a clumping of grain size, so that the size of the crystallite became large i.e. 54.58 nm.

Phases that appear at 10 h milling time are forsterite, silicon, and periclase. Crystalline size at 20 h milling time is 21.69 nm. While at the time of 40 h milling, the size of the crystallite is getting smaller to 18.78 nm. Changes in crystal size to milling time can be seen in **Figure 11**.

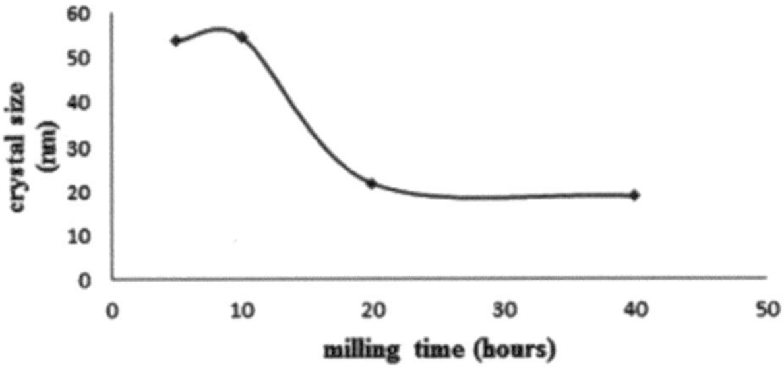

Figure 11.
Effect of milling time on crystal size.

Figure 12.
Differences in the form of morphology in each milling time variation with a magnification of 33,000×. (a) 5 h, (b) 10 h, (c) 20 h, (d) 40 h.

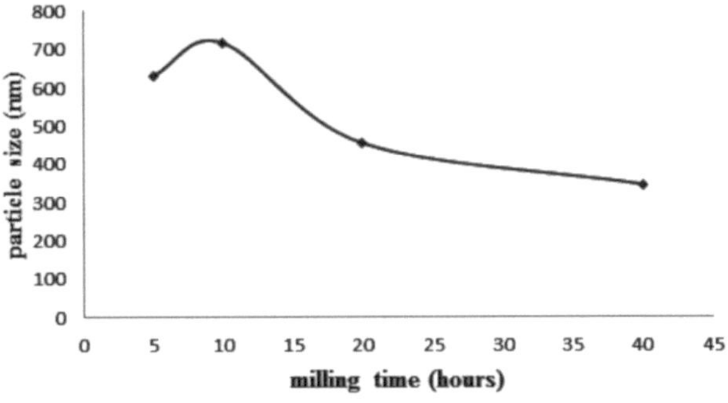

Figure 13.
Effect of milling time on grain size.

The milling time is very influential on the surface morphology of the sample, where the longer the process of milling is done, the smaller the particle size. **Figure 12** shows the morphology of forsterite powder with a variation of 5 h milling time. 10, 20 and 40 h were taken using SEM.

In **Figure 12**, there are differences in the form of forsterite morphology at each variation of milling time. At 5 h milling, the shape of the particles is round but not evenly distributed. Whereas at 10 h the particles clot back causing a large particle size. At 20 h the particle shrinks from the particle size to milling time of 10 h. The shape of the particles has seen a little round shape but has not been flat. At the time of milling 40 h the particle size has decreased and there are already smaller particles than before but the particles are not evenly sized. The grain size of forsterite particles for variations of milling time (a) 5 h, (b) 10 h, (c) 20 h, (d) 40 h are respectively 630, 717, 454, and 345 nm. The graph of changes in grain size to milling time can be seen in **Figure 13**.

9. High-energy milling methods, and the effect of the different parameters

The milling time parameter does not affect the crystal structure of forsterite, but it affects the appearance of other phases with different crystal structures. HEM-E3D is used for mechanical activation in samples using a speed of 500 rpm so that collisions occur between milling balls so as to produce energy. The impact energy is used to stretch or break the bonds of atoms in the sample. The longer the grinding time, the higher the temperature on the collision of the milling balls. In the process, the faster the ball mill rotation, the greater the energy produced and produces higher temperatures [29].

High temperatures benefit in some cases that require a diffusion process to support the integration process in the powder and reduce internal stress or even eliminate it. However, in some cases, the temperature increase is very detrimental because it can produce an unstable phase so that it will form other structures during the milling process and the powder size will be larger. At high temperatures, the crystal structure can grow quickly, but more defects in the crystals formed. The lattice parameters of the crystal will also change due to crystal defects by the collision. The material that has been heated is then cooled slowly so that the atoms in the material can be arranged regularly to occupy the lattices to form a crystal. At certain times there will be a phase change.

The number of crystalline phases has quite a combination of atoms or groups of atoms. The amorphous phase is relatively small because it does not have a long-range order and the atomic arrangement is not clear. The phase difference that occurs is inseparable from the influence of energy possessed by atoms for the diffusion process. At a certain level of energy, atoms can stay away from each other. If an atom has enough energy to break its bonds a diffusion process will occur [32].

Mechanical activation is related to the formation of material which causes strain on the solid mixture. This is one form of mechanical alloying, involving two constituent powders with heterogeneous size distributions that will affect the material properties and mechanism of phase formation of a material. The process looks simple, where different types of metals can bind through exchanging short distances between atoms called atomic diffusion, this can happen if atoms have enough energy to release bonds with the surrounding atoms so that they can move from the original lattice position to the empty lattice position [33].

The explanation of the reduction in crystal size due to the increase in milling time is that during the milling process of powder samples with variations in milling time there are four forces that occur in the material, namely impact (attrition), friction (shear), and compression (compression). With these four forces, the powder can be destroyed by a milling device in the formation of powder into nanoparticles. The size of the powder that has been crushed with a milling device becomes smaller than the size of the original powder.

The results of characterization using XRD obtained a relationship of 2θ with intensity. Every variation in the milling time shows that peaks appear and disappear and also the intensity decreases. The peak that disappears is because atoms in other phases do not exist so there is no scattering of atoms by certain structures. The measured intensity at XRD is the result of scattering intensity by certain atomic structures. The magnitude of the relative intensity of the series of peaks depends on the number of atoms or ions present and the distribution in the unit cell of the material. The decrease in intensity is due to changes in the size of the crystal, this change also causes the crystal structure of the forming element to change. The crystal size obtained when the 40-h milling time is 18.78 nm. For comparison the size of forsterite crystals obtained from talc $(Mg_3Si_4)_{10}(OH)_2$ and magnesium carbonate ($MgCO_3$) material is 33 nm [17].

When milling can measure the crystal grain size of the sample. This is because, during the milling process, the powder particles will experience repeated processes of destruction. When the milling balls collide with each other, the powder milled is between the collisions of the balls so that the powder will deform and the powder will disintegrate which will cause the grain size to be small and can also cause it to become large if the grain has clotted.

Compressive forces (compaction) that occur in particles other than destroying or breaking particles can also damage the pores on the surface of the particles, the pores are damaged due to the compressive force, especially small diameter pores that are very vulnerable to damage and disappear. In mills that are too long, particles can undergo agglomeration. After a long grinding and with very fine particles, the coupling forces become larger, and the presence of chemical bonds or Van Der Waals forces with bond strengths of 40–400 kJ/mol can make particles fuse or agglomerate. Or if there are particles trapped and then given an impact force, the particles can also be agglomerated. With the finer particles due to the long grinding time, the distance between particles will be smaller and more contact between particles will allow agglomeration to occur. Thus, on particles with porous surfaces, agglomeration allows for the formation of enlarged pore diameters due to "pore merging" due to agglomeration between particles.

Thus it can be explained by the increasing milling time used, the grain size of powder particles will be smaller. This is what allows the formation of nanometer scale powder particles due to the high collision frequency and the length of time the milling is used.

The grain size at the time of 10 h milling is 717 nm, there is a little clumping at 10 h milling time, while at 5 h 630 nm, when processed in 20 h the grain size decreases to 454 nm, at the time of milling 40 h the size of the crystal again decreased to 345 nm. The size of the grain increases during the milling time of 10 h because the particles agglomerate again after the particles are broken or broken by balls so that they are small and then group up to the milling time for 10 h this occurs due to the powder that has been solved and has a small effect. Unified powder causes large grain size, but at 20 h milling the grain size shrinks again because the sample experiences collision forces with balls that have high energy so the powder becomes eroded again and becomes small in size. In addition to the impact force, the powder also experiences other styles such as attrition, friction, and compression. So that at the time of milling 20 h forsterite has reached the fracture point and its size shrinks again.

The milling time also affects the sample morphology. Each milling variation will produce different morphological forms. The collision between the milling balls and the powder repeatedly over time increases the milling time will make the particle size of the powder will be smaller. The SEM results also show that the grain size is getting smoother as the milling time increases. This happens because in HEM-E3D there are collisions of milling balls with powder. Assault particles are trapped between steel balls that are colliding. Furthermore, powders undergo microscopic deformation and fracture processes (breaking) and welding (joining).

The forsterite particle size of the West Sumatra dunite has a grain size of 345 nm. The size obtained has not reached below 100 nm. The milling process carried out shows that forsterite powder has a tendency to agglomerate or clump, thus making forsterite under 100 nm not yet reached. This is due to the ratio of the ball to the powder used and also the speed of milling, and the length of time the milling is used. The higher the ball mill rotational speed will increase the collision and fineness of the powder, but if the speed has reached a certain level the fineness of the powder decreases. Likewise with the ratio of the weight of the milling ball to the powder, the variation in the number of pounding balls will also increase the collision contact area of the pounding and powder balls.

The difference in forsterite morphology is also seen in every variation of milling time, this is because the granules have clumped between the grains with each other, and the shape of the granules is round and uneven. Each increase in the variable milling time looks more like the morphology of the mixture, it appears that between the grains with each other have merged with each other, and the fusion indicates the formation of a solid solution on the powder. The increase in grain size in one of the milling results is due to the agglomeration process. The agglomeration process is the process of joining small particles into a larger structure through a physical binding mechanism.

The agglomeration process that occurs can also be caused by several things, namely sample powder contamination with crushing ball material and jar. Even though it has very high hardness, stainless steel on the crushing ball and jar will still give contamination to the ground powder sample. The high grinding speed and long grinding time cause contamination of the forming material of the crushing ball and the jar can be said to be almost unavoidable. Furthermore, which affects the grain size, namely the effect of the jar shape, the bottom edge design of HEM-E3D jar in the form of a curve can cause the formation of a dead zone which is an area where the powder is not grounded because the grinding media cannot reach it during milling.

10. Conclusions

In summary, has been described the synthesis and characterization of nanoparticles from forsterite (Mg_2SiO_4) sourced from dunite rocks in West Sumatra. Forsterite is obtained from dunite powder by giving calcination temperature. HEM-E3D is used to obtain nanoparticles from forsterite. The results show that milling time affects crystal size, grain size and morphology of forsterite. The higher the milling time, the smaller the particle size of forsterite. The forsterite particle size obtained has not reached below 100 nm. However, this result is better than previous research using the same method. This is because forsterite powder has a tendency for agglomeration.

Acknowledgements

The authors thank to RISTEK DIKTI for financial support through Hibah MP3EI 2016-2017 for this work.

Author details

Ratnawulan Ratnawulan* and Ahmad Fauzi
Padang State University, Padang, Indonesia

*Address all correspondence to: ratnawulan@fmipa.unp.ac.id

References

[1] Cooper R, Halle P. Reaction between synthetic mica and simple oxide compounds with application to oxidation resistant ceramic composites. Journal of the American Ceramic Society. 1993;**76**(5):1265-1273

[2] Sarimai S, Ratnawulan R, Ramli R, Fauzi A. Effect of milling time on particle size of forsterite (Mg_2SiO_4) from South Solok district. In: IOP Conference Series: Materials Science and Engineering. Vol. 335; Padang, Indonesia; 2018

[3] Thornhill M. Dry magnetic separation of olivine sand. In: Physicochemical Problems of Mineral Processing. Norway: Norwegian University of Science and Technology; 2011

[4] Rani AB, Annamalai AR, Majhi MR, Kumar AH. Synthesis and characterization of forsterite refractory by doping with kaolin. International Journal of ChemTech Research. 2014;**6**(2):1390-1397

[5] Saberi A, Alinejad B, Negahdari Z, Kazemi F, Almasi A. A novel method to low temperature synthesis of nanocrystalline forsterite. Materials Research Bulletin. 2007;**42**(4):666-673. DOI: 10.1016/j.materresbull.2006.07.020

[6] Mostafavi K, Ghahari M, Baghshashi S, Arabi AM. Synthesis of $Mg_2SiO_4:Eu^{3+}$ by combustion method and investigating its luminescence properties. Journal of Alloys and Compounds. 2013:**555**:62-67

[7] Mehta NS, Majhi MR. Effect of sintering at different temperature to enhance the physical and mechanical properties of forsterite refractories doped with kaolin. International Journal of Innovative Research in Science, Engineering and Technology. 2016;**5**(9):16261-16266. DOI: 10.15680/IJIRSET.2016.0509086

[8] Pawley R. The reaction talc+forsterite=enstatite +H_2O; new experimental results and petrological implications. American Mineralogist. 1998;**83**(1):51-57. DOI: 10.2138/am-1998-0105

[9] Aranovich L, Newton R. Experimental determination of CO2–H2O activity-composition relations at 600-1000°C and 6-14 k bar by reversed decarbonation and dehydration reactions. American Mineralogist. 1999;**84**(9):1319-1332. DOI: 10.2138/am-1999-0908

[10] Maliavski N, Dushkin O, Markina J. Forsterite powder prepared from water soluble hybrid precursor. AICHE Journal. 1997;**43**(11A):2832-2836. DOI: 10.1002/aic.690431301

[11] Shieh Y, Rawlings R, West D. Constitution of laser melted alumina-magnesia-silica ceramics. Materials Science and Technology. 1995;**11**(9):863-869. DOI: 10.1179/mst.1995.11.9.863

[12] Jeevanandam J, Barhoum A, Chan YS, Dufresne A, Michael K, Danquah MK. Review on nanoparticles and nanostructured materials: History, sources, toxicity and regulations. Beilstein Journal of Nanotechnology. 2018;**9**:1050-1074. DOI: 10.3762/bjnano.9.98

[13] Bell TE. Understanding Risk Assesment of Nanotechology. USA: Artikel National Nanotechnology Coordination Office; 2006

[14] Kishore K. Study on the effect of high energy ball milling (a nano material process) on the microstructure and mechanical properties of a (Al-Si-Fe-Cu) alloy [Thesis]. Rourkela: National Institute of Technology; 2007

[15] CSSR K. Nanofabrication Towards Biomedical Applications. Weinheim, Germany: Wiley-VCH Verlag GmbH & Co. KGaA; 2005

[16] Tavangarian R, Emadi R. Mechanical activation assisted synthesis of pure nanocrystalline forsterite powder. Journal of Alloys and Compounds. 2009;**485**(1-2):648-652

[17] Tavangarian R, Emadi R. Mechanochemical synthesis of single phase nanocrystalline forterite powder. International Journal of Modern Physics B. 2010;**24**(3):343-350

[18] Cheng L, Liu P, Chen XM, Niu WC, Yao GG, Liu C, et al. Fabrication of nanopowders by high energy ball milling and low temperature sintering of Mg_2SiO_4 microwave dielectrics. Journal of Alloys and Compounds. 2012;**513**:373-377. DOI: 10.1016/j.jallcom.2011.10.051

[19] Koukouzasa N, Gemeni V, Ziock HJ. Sequestration of CO2 in magnesium silicates, in Western Macedonia, Greece. International Journal of Mineral Processing. 2009;**93**(2):179-186. DOI: 10.1016/j.minpro.2009.07.013

[20] Abdel-Karim AAM, Elwan WI, Helmy H, El-Shafey SA. Spinels, Fe-Ti oxide minerals, apatites, and carbonates hosted in the ophiolites of Eastern Desert of Egypt: Mineralogy and chemical aspects. Arabian Journal of Geosciences. 2014;**7**(2):693-709. DOI: 10.1007/s12517-013-0854-0

[21] Mohanty JK. Characterization of high magnesian rocks for suitability as flux in iron and steel industry. Journal of Geology and Mining Research. 2009, 2009;**1**(7):149-155. DOI: 10.5897/JGMR

[22] Nelson A. Silicate structures, structural formula, neso-, cyclo-, and soro-silicates. In: Mineralogy. New Orleans: Louisiana Tulane University; 2012

[23] Perepelitsyn VA, Kuz'menko NG, Kuperman YE, Visloguzova EA. The mineral composition of the dunites from the Solov'evogory deposits. Refractories. 1974;**15**(1-2):92-97. DOI: 10.1007/BF01286316

[24] Didem A. Petrogenesis of subduction zone and dunite bodies. International Journal of Earth Sciences and Engineering. 2012;**2**:377-386

[25] Quinn AW. Dana's manual of mineralogy (Hurlbut, Jr.; Cornelius S.). Journal of Chemical Education. 1952;**29**(10):532. DOI: 10.1021/ed029p532.2

[26] Ray W, Kent RW, Simon P, Kelley SP, Malcolm S, Pringle MS. Mineralogy and 40Ar/39Ar geochronology of orangeites (group II kimberlites) from the Damodar Valley, Eastern India. Mineralogical Magazine. 1998;**62**(3):313 323. DOI: 10.1180/002646198547701

[27] Kosanovic C. Synthesis of forsterite powder from zeolite precursors. Croatica Chemica Acta. 2005, 2005;**2**(79):203-208

[28] Decastro C, Mitchell B. Nanoparticles from mechanical attrition. In: Baraton M, Valencia CA, editors. Synthesis, Functionalization and Surface Treatment of Nanoparticles. Valencia: American Scientific Publishers; 2002. pp. 1-15

[29] Calka A, Nikolov JI. The dynamics of magneto-ball milling and its effects on phase transformations during mechanical alloying. Materials Science Forum. 1995;**179-181**:333-338. DOI: 10.4028/www.scientific.net/MSF.179-181.333

[30] Ozkaya T, Toprak MS, Baykal MSA, Kavas AH, Koseoglu HY, Aktas B. Synthesis of Fe_3O_4 nanoparticles at 1000C and its magnetic characterization. Journal of

Alloys and Compounds. 2008;**472**:18-23. DOI: 10.1016/j/jallcom.2008.04.101

[31] Riazian M, Bahari A. Structure of lattice strain and effect of sol concentration on the characterization of TiO_2-CuO-SiO_2 nanoparticles. International Journal of Nano Dimension. 2012;**3**(2):127-139

[32] Van V. Elements of Materials Science and Engineering. Publication Reading, MA: Addison-Wesley; 1989

[33] Eckert J, Holzer JC, Krill CE, Johnson WL. Structural and thermodynamic properties of nanocrystalline fcc metals prepared by mechanical attrition. Journal of Materials Research. 1992;**7**(7):1751-1761. DOI: 10.1557/JMR.1992.1751